DERIVE® LABORATORY MANUAL

FOR

Differential Equations

DERIVE® LABORATORY MANUAL

FOR

Differential Equations

DAVID C. ARNEY

United States Military Academy

ADDISON-WESLEY PUBLISHING COMPANY

Reading, Massachusetts • Menlo Park, California • New York
Don Mills, Ontario • Wokingham, England • Amsterdam • Bonn
Sydney • Singapore • Tokyo • Madrid • San Juan • Milan • Paris

Reproduced by Addison-Wesley from camera-ready copy supplied by the author.

ISBN 0-201-57886-7

1 2 3 4 5 6 7 8 9 10-BK-9594939291

Preface

Today we render unto the computer what is the computer's, and unto analysis what is analysis', we can think in terms of general principles, and appraise methods in terms of how they work ...

— Peter Lax [1989]

This book is designed to show the reader how to use the software package Derive®[1] to help solve problems from differential equations courses. It is a companion manual for any of the textbooks used in these courses and also could be used in other courses covering this topic (for example: engineering mathematics and applied mathematics).

Derive is a computer algebra system (CAS) and in capable hands can be a powerful tool in problem solving. It is mostly menu driven, making it user-friendly and easy to use. Its screen displays are nicely formatted and easy to read. Derive provides capabilities to perform symbolic, graphic, and numeric manipulation.

This manual is organized into three chapters. The first chapter contains a short, general explanation of the basic capabilities and limitations of Derive. This chapter generally covers the symbolic, graphic, and numeric capabilities of the software in the areas of algebra, calculus, matrix algebra, complex variables, and special functions. The basic keystrokes, menu commands, mode selections, plotting parameters, and screen displays are presented. The second chapter contains examples of solutions to problems similar to those found in differential equations textbooks using Derive to perform some of the manipulation, plotting, and analysis. Some of the problems are models from applications in the various engineering disciplines (mechanical, electrical, systems, and civil); science; and economics. Other problems deal directly with the mathematical concepts of topics in differential equations. Each example is designed in such a manner that the

[1] Derive® is a registered trademark of Soft Warehouse, Inc., Honolulu, HI

student should be able to work through it while seated at a computer with this manual open. The third chapter includes laboratory exercises which require the use of Derive to solve, explore, and analyze problems. Some of these problems are similar to those in the second chapter, while others introduce new concepts in both the mathematics and the software. These laboratory exercises lead the reader through the steps of problem solving. Sometimes the exercises involve conducting a computational experiment to find an answer, perform an analysis, or determine a value for a parameter. Students should work these laboratory exercises and experiments to develop their problem-solving skills.

In writing this manual, I have assumed that the reader is familiar with differential equations and their solutions. The object of this book is not to teach the solution techniques, but to show how they are used in the efficient and powerful way provided by the software to solve and analyze problems. Since the manual is designed for undergraduate students, only elementary equations and techniques are discussed.

It is hoped that this guide can contribute to better student understanding and performance in differential equations problem solving. It is also hoped that it illustrates the power of Derive as a computational tool that enables students to become better problem-solvers.

Many people have contributed to the creation of this book. I thank them all. I give special thanks to Frank Giordano for his encouragement; Jack Robertson, Fred Rickey, and Joe Myers for their careful review; David Stoutemyer and Albert Rich for their excellent software; and David Pallai and Angie Davis for their editorial assistance. Finally, I thank my wife, Sue, and my children, Kristin, Dan, Lisa, and Katie for their patience and understanding.

Chris Arney
West Point, NY
25 August 1990

Contents

Notation and Technical Information

The best way ...to make mathematics interesting to students ...is to approach it in a spirit of play.

— Martin Gardner [1975]

This manual was typeset using LaTeX. This text processing system allows for special notation and fonts to be used to help identify the computer keystrokes and input commands. Boldface type is used to indicate the precise key on the keyboard to strike. This is especially helpful for identifying the special keys like **Enter**, **Ctrl**, **Alt**, and the function keys (**F1, F2, ..., F10**). The **Ctrl**, **Alt**, and **Shift** keys are used in combination with another key and this is denoted by showing both keys in boldface (i.e. **Alt-e** or **Ctrl-Enter**).

The teletype font is used for the commands and menu selections for input to or output from the software (i.e. `Author`, `Simplify`, `Help`, `EXP(z)`, `SIN(2xy)`). The inline commands are in upper case to distinguish them from the menu selections and are displayed in a box for further emphasis when they are in the exact form to be entered into the computer. Derive is not sensitive to case, so actual input may be made with upper- or lower-case characters.

Sometimes italics are used to indicate a pseudo-command. For example, the selection of the `Author` menu command and the input of an expression can be denoted by *Authoring* the expression. This notation is not used often and only after the command is very familiar to the reader.

Words in uppercase letters and in regular typeface are used to indicate DOS file names. Derive uses files to store extra commands and functions, system states, and output expressions.

Version 1.60 of Derive was used in all the examples in this book. Earlier versions of the software lacked the two utility files ODE1.MTH and

ODE2.MTH which play a significant role in solving differential equations. Of course, later versions of the software may provide more commands and added capabilities which are not discussed or used in this manual.

There are numerous figures throughout the manual showing actual screen images from Derive. Usually, the highlight is moved off the visual region of the screen so it will not interfere with the display. The screen images were produced using the CAPTURE.COM program of Microsoft Word and plotted in Postscript format on an QMS Laser Printer. The computer that was used to run the software was a Zenith 248Z, which is compatible with an IBM-AT, with a VGA card. Derive was run in its VGA mode.

DERIVE® LABORATORY MANUAL

FOR

Differential Equations

Chapter 1

Fundamentals

The computer is no better than its program.

—Elting Elmore Morison [1966]

This chapter describes the fundamentals of using Derive to solve mathematical problems. After understanding the basics presented here, you will find yourself well-prepared to understand the solved example problems in Chapter 2. Since differential equations involve many areas of mathematics, sections are included on the use of Derive in the subjects of calculus, linear algebra, and complex variables. This chapter also includes fundamental concepts in symbolic algebra, plotting, and numerical approximations. However, neither this first chapter nor this manual are complete reference manuals for Derive. They are not intended to replace the *Derive User Manual* that comes with the software. In fact, this manual will refer to the *Derive User Manual* at times.

Beginning users of Derive should definitely spend some time in this chapter before tackling the examples in Chapter 2. The material presented here is good, helpful reference for understanding the commands used in Chapter 2, for solving the problems in Chapter 3, or for using Derive to solve problems in many areas of mathematics.

1.1 An Overview of Derive

Making mathematics more exciting and enjoyable is the driving force behind the development of Derive. The system is destined to eliminate the drudgery of performing long tedious mathematical calculations.

—*Derive User Manual* [1990]

It is important to realize that the Derive software is not specially designed to solve any one kind of problem (for example: differential equations, matrix algebra, or calculus). There are other packages like Phaser[1], MDEP[2], MacMath[3] and Differential Equations Graphics Package[4] that are designed for the express purpose of solving differential equations, and other packages like LinTek[5], MAX[6], and MATLAB[7] that primarily solve linear algebra problems. Some of the software packages that help solve calculus problems are Calculus Toolkit[8], CALCULUS PAD[9], Calculus[10], Mathematics Plotting Package[11], Exploring Calculus[12], and MathCAD[13]. These calculus packages possess some of the capabilities of Derive, but they lack its versatility. Most of the differential equations packages use numerical methods to approximate the solutions and have special graphics available to plot solutions, direction fields, trajectories, and phase planes. On the other hand, Derive is a computer algebra system (CAS) whose function is more general than any one subject or topic. In fact, its primary function is not to solve differential equations; instead, it performs symbolic manipulation and numerical computation that are used for many of the mathematical steps and approximations involving algebra, calculus, trigonometry, and plotting necessary to solve, analyze, and study differential equations and many other types of problems. Other CAS packages similar to Derive for personal com-

[1] H. Kocak, Springer-Verlag
[2] J.L. Buchanan, United States Naval Academy
[3] J.H. Hubbard and B.H. West, MacMath
[4] Sheldon Gordon, MatheGraphics
[5] John Fraleigh, Addison-Wesley
[6] E.A. Herman and C.H. Jepsen, Brooks-Cole
[7] The Math Works
[8] Ross Finney, et.al., Addison-Wesley
[9] I. Bell, et.al., Brooks-Cole
[10] John Kemeny, TrueBASIC
[11] H. Penn, United States Naval Academy
[12] John Fraleigh and Lewis Pakula, Addison-Wesley
[13] Mathsoft; student version available from Addison-Wesley

puters include Maple[14], Mathematica[15], muMath[16], Macsyma[17], Reduce[18], and Theorist[19].

The following figure is a sample Derive screen from a problem-solving session showing the general layout of Derive's user interface.

```
1:   COS (x)

        π
     ⌠
2:   ⌡   COS (x) dx
        0

3:   █
________________________________________________________________
COMMAND: Author Build Calculus Declare Expand Factor Help Jump soLve Manage
         Options Plot Quit Remove Simplify Transfer moVe Window approX
Enter option
Simp(2)                               Free:100%              Derive Algebra
```

Derive Screen showing three working expressions in the work area and the main menu

We will explain how expressions like these are created later in this chapter and throughout Chapter 2.

The display screen is organized into several sections each with an important function. The top part contains the working expressions. The following sample of this section of the Derive screen shows expressions numbered 5-7.

```
         3
5:    x   - 2 x + 4

6:    (x - 1 - î) (x - 1 + î) (x + 2)

                  2
7:    (x + 2) (x   - 2 x + 2)
```

Three algebraic working expressions (numbered 5-7) in the top part of a Derive screen.

This section can also include plots and can be windowed into subsections by the user. The next screen display shows three windows, one containing a 3-dimensional plot, another window containing a 2-dimensional plot, and the third has algebraic expressions. Windowing and plotting are discussed in Sections 1.4 and 1.10, respectively.

[14] Waterloo Maple Software
[15] Wolfram Research
[16] Soft Warehouse
[17] Symbolics
[18] Northwest Computer Algorithms
[19] Prescience

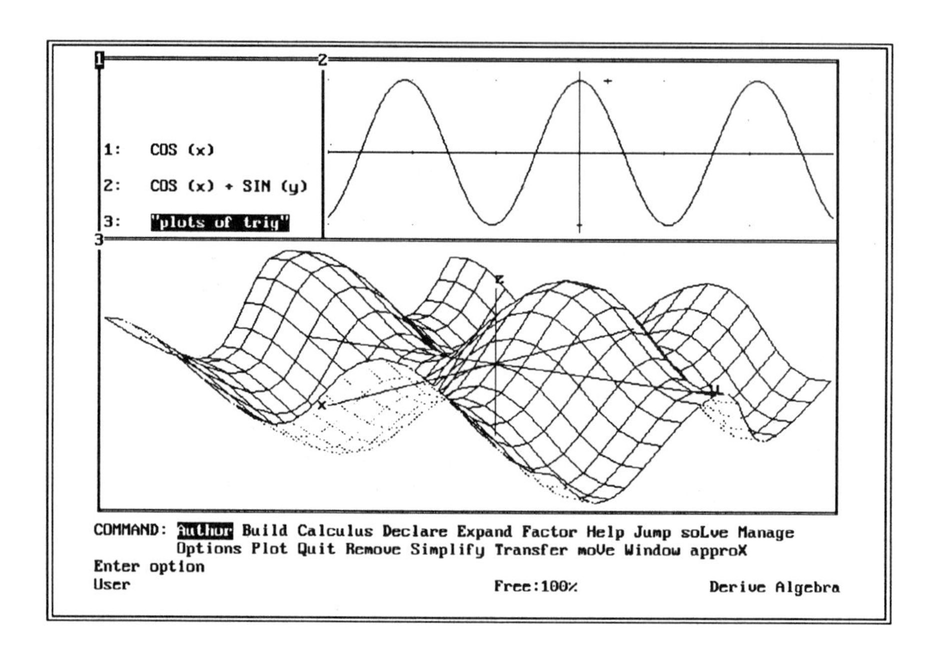

Derive screen with a 3-dimensional plot, a 2-dimensional plot, and algebraic expressions.

1.2 Menus and Commands

Mathematics, rightly viewed, possesses not only truth, but extreme beauty ...

— Bertrand Russell [1902]

The section below the working expressions on the Derive screen contains the menu. Commands are selected from the menu in one of two ways. The first way is merely to type the letter associated with the command (the upper case letter in the display of the command, which is not always the first letter of the word). The second way is to move the highlight using **Space, Tab, Backspace,** or **Shift-Backspace** and pressing the **Enter** key when the desired command is highlighted. The **Esc** key is used to abort a command or menu without making any changes. The selections in the main menu, especially the **Author** command, are the user's usual interfaces with Derive.

```
COMMAND: Author Build Calculus Declare Expand Factor Help Jump soLve Manage
         Options Plot Quit Remove Simplify Transfer moVe Window approX
Enter option
                                     Free:100%              Derive Algebra
```

Derive's familiar main menu and its command selections

Most of the commands are executed directly from the main menu. The executable commands are **Author, Build, Expand, Factor, Help, Jump, soLve, Quit, Remove, Simplify,** and **moVe**. The following table gives a short description of each of these executable main menu commands.

Command	Function
Author	Allows the input of expressions into the algebra window or work area
Build	Allows the building of an expression from previous expressions
Expand	Performs an algebraic expansion of an expression
Factor	Factors the expression; if the expression is a polynomial it will attempt to factor it into the lowest order terms meeting the criteria established with the mode portion of the command; if the expression is a number, it will express it in terms of its prime divisors
Help	Allows user to get help on the function of a command (see Section 1.5)
Jump	Moves the highlight to a given expression number; a good way to move to an earlier expression in the work area
soLve	Solves for the desired variable in an equation
Quit	Stops the Derive program
Remove	Removes expressions from the work area
Simplify	Frequently used command for simplifying an expression, equation, or a Derive command in the work area
moVe	Rearranges expressions in the work area

Main menu commands and their functions

The nonexecutable main menu commands provide submenus for additional commands. One such submenu selection which will be very useful for our purposes is **Calculus**. The commands in this submenu are **Differentiation, Integration, Limit, Product, Sum,** and **Taylor**. These calculus menu commands also have equivalent in-line commands. The next table provides a short description of the functions of the menu commands in the **Calculus** submenu.

Command	Function
`Differentiation`	Finds derivatives (regular or partial) of any integer order
`Integration`	Finds either the antiderivative or the definite integral, depending on whether or not limits of integration are entered
`Limit`	Finds the limit of an expression as one of the variables approaches a given point
`Product`	Finds definite product or antiquotient of an expression (index variable and upper and lower limits are required)
`Sum`	Finds the definite sum or antidifference of an expression (index variable and upper and lower limits are required)
`Taylor`	Finds a Taylor polynomial approximation to an expression (an expansion variable, expansion point, and order of the polynomial are required)

`Calculus` menu commands and their functions

The equivalent in-line commands to perform these calculus functions are presented in the following table. The in-line commands must be entered along with their arguments using the `Author` menu command; the `Simplify` command then executes the operation.

Command	Function
DIF(u,x,n)	Finds the $n^{th} - orderderivativeofuwith$ respect to x
INT(u,x)	Finds the antiderivative of u with respect to x
INT(u,x,a,b)	Finds the definite integral of u with respect to x from a to b
LIM(u,x,a)	Finds the limit of u as x approaches a from above
LIM(u,x,a,0)	Finds the limit of u as x approaches a from below
PRODUCT(u,n)	Finds the antiquotient of u with respect to n
PRODUCT(u,n,k,m)	Finds the definite product of u as n goes from k to m
SUM(u,n)	Finds the antidifference of u with respect to n
SUM(u,n,k,m)	Finds the definite sum of u as n goes from k to m
TAYLOR(u,x,a,n)	Finds the Taylor polynomial approximation of order n to u about the point $x = a$

Calculus in-line commands and their functions

The set of commands in the submenu **Declare** are **Constant, Function, Variable, Matrix,** and **vectoR**. These are very useful and powerful commands used to establish mathematical structures. Careful thought in their use is often needed to solve problems correctly. These commands will be used in some of the example problems and may be necessary in some of the laboratory exercises. Not only can you declare these mathematical structures, but also important range and domain information on variables is established using the **Declare, Variable** command. This table describes the functions of these commands:

Command	Function
Constant	Declares a constant and can assign it a value
Function	Declares and defines a function
Variable	Assigns the domain of a variable (options include **Positive, Nonnegative, Real, Complex,** and **Interval**)
Matrix	Establishes the dimensions of a matrix and provides entry of elements for the matrix
vectoR	Establishes the dimension of a vector and provides entry of the elements for the vector

Declare menu commands and their functions

The **Manage** submenu contains the commands **Branch, Exponential, Logarithm, Ordering, Substitute,** and **Trigonometry**. The most important and frequently used command in this list is the **Manage, Substitute** command. This command substitutes values for variables or subexpressions of an expression.

The **Options** submenu sets up different modes of operation for Derive. Some of the **Options** commands are used before any evaluations are done. Other commands may be performed during the course of problem solving. The commands in this submenu are **Color, Display, Execute, Input, Mute, Notation, Precision,** and **Radix**. The following table provides a short description of these commands. It is important to set the proper parameters for your computer using the **Options, Display** command. These settings can be saved via the **Transfer, Save, State** command.

Command	Function
Color	Used to change the color of the work area, menu, or plotting curves
Display	Used to set **Mode** (**Text** or **Graphics**), **Resolution**, and **Adapter**
Execute	Allows the execution of a DOS command
Input	Establishes variables as single letters or words
Mute	Sets the error beep on or off
Notation	Sets the style of notation for numerical output
Precision	Sets computation precision and the digits of accuracy for the **Approximate** mode
Radix	Sets the radix base for number input and output

Options menu commands and their functions

The **Transfer** submenu provides for input and output in Derive. The subcommands and their functions are given in the following table:

Command	Function
Merge	Adds expressions stored in .MTH file to those in the working area
Clear	Deletes all the expressions from the window
Demo	Loads and simplifies expressions from a .MTH file one at a time
Load	Loads the expression in a .MTH file into the working area
Save	Saves the expressions in the working area to a file; options include Derive (.MTH), FORTRAN, Pascal, or Basic format
Print	Prints expressions to a printer or file; allows for the setting of several printing options
sTate	Saves the current state of Derive parameters to a .INI file for reloading or loads a previously saved (See Sections 1.5 and 1.14 for more information)

Transfer menu commands and their functions

There are some additional commands (usually in the form of functional operators or functions) in Derive that are not reached through any menu. These commands and functions must be typed into the workspace using the **Author** command. Some of the more useful and common commands not in a menu are **EIGENVALUES**, **DET**, **DIMENSION**, **CURL**, **FIT**, **GRAD**, **LAPLACIAN**, **VECTOR_POTENTIAL**, **STDEV**, **RMS**, **ROW_REDUCE**, **PRODUCT**, **POTENTIAL**, **PHASE**, **IM**, and **REAL**. Explanations of these operations and others are found in the *Derive User Manual* and many of them are described later in this chapter. Some of these operations will be used in the example problems of Chapter 2.

1.3 Keystrokes

There are things which seem incredible to most men who have not studied mathematics.

—Archimedes

As demonstrated in Section 1.2, the keystrokes used to select menu commands in Derive are straightforward. Just move the highlight by hitting the space bar and pressing **Return** or **Enter** when the desired command is highlighted, or merely type the letter that is capitalized in the desired command. However, there are many other keystrokes to learn in order to enter and edit expressions when using the **Author** and other similar commands. Keyboard skill is essential and a prerequisite for efficient use of the software. Learning to use the editing features can considerably reduce the amount of retyping of expressions.

One important feature is the ability to input comments into the work area between expressions. Comments can be entered by putting double quotes (") around the line of input when entering an expression using the **Author** command.

When entering and editing an expression using the **Author** command, the **Backspace** deletes the previous character (to the left of the cursor). **Ctrl-S** moves the cursor one character to the left without erasing the character, and **Ctrl-D** moves the cursor to the right. **Ctrl-A** moves the cursor a group of characters (or token) to the left, and **Ctrl-F** moves the cursor a token to the right. **Ctrl-Q-S** moves the cursor to the left end of the line, and **Ctrl-Q-D** moves the cursor to the far right end. **Del** deletes the character at the cursor. **Ctrl-T** deletes a token at a time, and **Ctrl-Y** deletes the entire line.

Ins toggles the typing modes between inserting between existing characters and overwriting the existing characters. When the insert typing mode is in effect, the word **Insert** appears on the status line below the menu.

The direction keys are used to highlight and extract subexpressions (or entire expressions) from previous expressions in the working area. These subexpressions can be expanded or assembled into new expressions. The ↑ and ↓ move the highlight one expression at a time. **PgUp** and **PgDn** move the highlight several expressions at once. **Home** moves to the first expression, and **End** highlights the last expression in the working area. The other direction keys, ← and →, move the highlight through the individual tokens (or subexpressions) of an expression. Even finer highlighting is possible (see the *Derive User Manual*). The **F3** function key inserts whatever is highlighted into the author line; the **F4** key does the same, but automatically places parentheses around the expression. Similar manipulations can be performed using the `Build` command (Section 1.2). The easiest way to refer to previous expressions is by their leading reference number in the work area. Just use the # symbol preceding the number of the expression to obtain that expression. For example, authoring **#4/#6**, inserts expression number 4 divided by expression number 6 into the work area.

The **Enter** key completes the authoring of an expression and adds it into the work area. **Ctrl-Enter** also does this along with automatically simplifying the expression.

There are two modes for variable names—character or word. The mode is set with the **Options, Input** command. Also, Greek letters can be used as variables and entered using the **Alt** key. Greek letters are produced by holding down the **Alt** key and entering the corresponding English letter. See the *Derive User Manual* for more details and for a table of available Greek letters.

1.4 Screen Displays and Windows

He cannot know that he likes raspberry pie if he has never tasted raspberry pie.

—George Polya [1946]

Derive allows the display screen to contain more than one window. These windows can split the screen in a variety of ways or can overlay one another. Windowing can be used for several purposes. Separate windows can contain solutions to different problems, different developments of the same problem, different perspectives or scales of the same plot, or maintain important results in one window while working in another. Windows can be especially helpful to show symbolic computations, numeric computations, and graphics all on the same screen in three or more different windows. The commands in a work area of one window are available for use in another window. This is very handy when using commands loaded from a utility file (see Section 1.12 for an explanation of utility files).

The **Window** sub-menu is called from the main menu. The **Window** menu is shown here:

```
WINDOW: Close Designate Flip Goto Next Open Previous Split
Enter option
                                   Free:100%          Derive Algebra
```

Be sure to use windowing when it can be helpful. Windowing is a very important and powerful feature of Derive.

The following table describes the use of these **Window** menu commands.

Command	Function
`Close`	Closes the active window
`Designate`	Designates the active window as one of 3 types (Algebra, 2D plot, 3D plot)
`Flip`	Flips or rotates between overlaid windows (not numbered separately)
`Goto`	Activates and moves to the designated window
`Next`	Activates and moves to the next window in sequence
`Open`	Opens a new window and designates it as one of 3 types (Algebra, 2D plot, 3D plot)
`Previous`	Moves to the previous window in sequence
`Split`	Splits the current active window to form a new window; split can be made vertically or horizontally and can be established in different sizes

`Window` menu commands and their functions

The following example shows three horizontal windows. The top window contains the symbolic form of the function, the middle window shows the global behavior of this function, and the bottom window gives a smaller scale perspective near the origin.

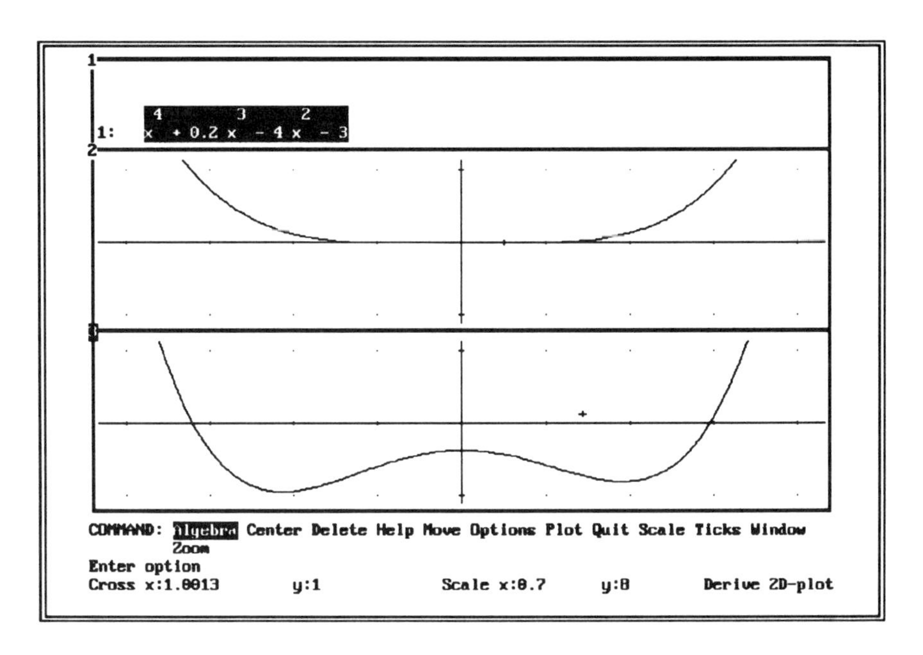

There is a convenient keystroke that saves time when moving between windows. The **F1** key provides easy and rapid movement through the existing windows.

1.5 On-Line Help and System State

If I should not be learning now, when should I be?

—Lacydes [241 B.C.]

The **Help** command displays the following menu:

```
                    Derive Help Menu

              E - line Editing commands
              F - Functions and constants
              A - Algebra window commands
              2 - 2D-plot window commands
              3 - 3D-plot window commands
              S - current State of system
              R - Return to Derive

           Select one of the above subjects

HELP: Editing Functions Algebra 2D-plot 3D-plot State Resume

Enter option
                              Free:100%            Derive Algebra
```

To obtain on-screen help on any of these subjects, press the appropriate letter in this menu. Helpful information is provided along with the page number in the *Derive User Manual* which contains more information. There is plenty of information available, so don't be hesitant to use this feature.

The **Help, S** option displays the current system state. There are two screens of information available through this command. The first screen provides the basic mode settings and the second screen contains mostly information on the plotting parameters and output formats. Unfortunately, no information is provided about the status of the display modes set with the **Options, Display** menu commands (**Text** or **Graphics**, resolution, type of graphics adapter–**CGA, EGA, VGA**, etc.). However, all of these modes are saved and established by the .INI file. The system status for the set- up used to solve the examples and exercises in this manual are provided in the following output.

```
                              Current State of System

   Precision mode:  Exact
   Precision digits:  6
   Notation style:  Rational
   Notation digits:  6
   Input radix base:  10
   Output radix base:  10
   Input mode:  Character

   Branch selection:  Principal
   Logarithms:  Collect
   Exponentials:  Collect
   Trig functions:  Auto
   Trig powers toward:  Auto

   Rows per tick mark:  4
   Columns per tick mark:  9
   2D plotting accuracy:  8
   2D plot mode:  Rectangular
   Axes color:  15
   Cross color:  15
   3D top color:  6
   3D bottom color:  5
   ________________________________________________________________________
   HELP: Next Previous Resume

   Enter option
                                        Free:100%              Derive Algebra
```

```
                              Current State of System

   Frame color:  15
   Option color:  15
   Prompt color:  15
   Status color:  15
   Menu background color:  0
   Border color:  0

   Work color:  15
   Work background color:  0
   Video mode:  18

   Page length:  66
   Page width:  80
   Top margin:  0
   Bottom margin:  0
   Left margin:  8
   Right margin:  3
   ________________________________________________________________________
   HELP: Next Previous Resume

   Enter option
                                        Free:100%              Derive Algebra
```

These state parameters can be saved into and reloaded from an .INI file using the **Transfer, sTate** commands. Derive starts in the state set in the file DERIVE.INI. If a different initial state is desired, save that state in the file DERIVE.INI and that will become the initial state of the software. Different states can be saved in other .INI files and loaded whenever needed. See Section 1.2 or the *Derive User Manual* for more information about the system state and refer to Section 1.14 for recommendations on establishing proper mode settings for your own DERIVE.INI file.

1.6 Symbolic Algebra

Angling may be said to be so like the mathematics that it can never be fully learnt.

— Izaak Walton [1653]

Derive has the capability to do many algebraic manipulations using the **Simplify, soLve, Manage, Expand,** and **Factor** commands. See Section 1.2 or the *Derive User Manual* for explanation of these commands. Derive can work with equations, expressions, and inequalities. It also can do operations involving ∞ (entered by typing **inf**).

Derive assumes the default declaration of **Real** for all variables, unless over-ridden by the **Declare, Variable** command. In a similar manner the **Declare, Constant** command creates user-defined constants. The constant number e is entered by typing **Alt-e** and displayed as ê, and the constant π is entered by **pi** or **Alt-p**. Three utility files are available to the Derive user with many useful constants already declared. These files are ENGLISH.MTH, METRIC.MTH, and PHYSICAL.MTH and can be loaded when needed using the **Transfer, Load** command.

Declare, Function is a very powerful feature in Derive. It creates user-defined functions which also can be used as functional operators. See Example 2.12 for use of this command in solving differential equations and in finding Laplace transforms. However, do not over-use this feature. You can work with equations without declaring them as functions.

Some of the common algebraic and trigonometric functions in Derive are provided in the following table:

Command	Function
EXP(z)	Exponential of z, displayed as $\hat{e}^z$
SQRT(z)	$z^{1/2}$, takes into account the domain of the variable and the branch
LN(z), LOG(z)	Natural logarithm and principal natural logarithm
LOG(z,w)	Logarithm of z to base w
PI, DEG	Constants π and $\pi/180$
SIN(z), COS(z), TAN(z) COT(z), SEC(z), CSC(z)	Trigonometric functions
ASIN(z),ACOS(z),ATAN(z) ACOT(z),ASEC(z),ACSC(z)	Inverse trigonometric functions
SINH(z),COSH(z),TANH(z) COTH(z),SECH(z),CSCH(z)	Hyperbolic trigonometric functions (Derive simplifies to an exponential equivalent)
ASINH(z),ACOSH(z),ATANH(z) ACOTH(z),ASECH(z),ACSCH(z)	Inverse hyperbolic trig functions (Derive simplifies to logarithm equivalent)
ABS(x), SIGN(x)	$\|x\|$ and sign of x, respectively
MAX($x_1, ..., x_n$),MIN($x_1, ..., x_n$)	Max and min over the set of arguments
STEP(x)	1 if $x > 0$, 0 if $x < 0$

Common algebraic and trigonometric functions

Trigonometric manipulation is controlled in Derive with the **Manage, Trigonometry** commands. There are many possible combinations of ways to manage and to simplify the trigonometric functions. The **Manage, Trigonometry** submenu contains settings for **Direction**: **Auto**, **Collect**, or **Expand** and for **Toward**: **Auto**, **Sine**, and **Cosine**. See the *Derive User Manual* for help, or try the different combinations of these settings on simple expressions. Derive uses radians as the unit of angular measure in all its trigonometric functions.

There are many other functions in Derive that can be used to perform operations in probability, statistics, and finance. None of these are described in this manual. See the *Derive User Manual* for descriptions of these special functions.

1.7 Calculus

What DesCartes did was a good step. ...If I have seen further it is by standing on ye sholders of Giants.

—Issac Newton [1676]

Derive performs the calculus operations of limits, differentiation, antidifferentiation, integration, Taylor polynomial approximations, series summation, and products. These operations can be done through the **Calculus** menu or with in-line commands. Several items about these operations are worth mentioning: i) antidifferentiation does not automatically produce an arbitrary constant, ii) integration in the **Approximate** mode is performed with Simpson's adaptive quadrature, and iii) improper integrals are evaluated as Cauchy Principal Values. See Section 1.2 and the *Derive User Manual* for explanations of these Calculus commands and their options.

There are three utility files that contain additional calculus-related functions: DERIV.MTH, INTEGRAL.MTH, and SPECIAL.MTH. The file INTEGRAL.MTH contains functions for arc length, volume of revolutions, surface area, Fourier series approximations (see Sections 2.11 and 3.17), and Laplace transforms (see Sections 2.12 and 3.18).

Derive performs several vector calculus operations. The default representation for the vector operators is rectangular coordinates. However, the utility file COORD.MTH can help perform coordinate transformations and produce operations in other coordinate systems. See the *Derive User Manual* for the description of this file and its use. The following table provides descriptions of some of the common vector calculus commands:

Command	Function
`GRAD(f(x,y,z))`	Finds the gradient of an expression
`DIV([u(x,y,z),v,w])`	Finds the divergence of a vector
`LAPLACIAN(f(x,y,z))`	Finds the divergence of the gradient of an expression
`CURL([u,v,w])`	Computes the curl of a vector
`POTENTIAL([u,v,w])`	Calculates the scalar potential of the given vector (if it exists)
`VECTOR_POTENTIAL([u,v,w])`	Calculates the vector potential of the given vector (if it exists)
`JACOBIAN([u,v,w],[r,s,t])` (in COORD.MTH)	Produces the Jacobian matrix of a transformation, where u, v, w are functions of r, s, t

Common vector calculus operations

1.8 Matrix Algebra

I believe we should go to the moon ... before the decade is out.

—John F. Kennedy [1961]

Solving systems of linear equations and performing matrix and vector operations are important and frequent tasks in solving problems involving systems of differential equations. Derive has capabilities to help the user perform these operations.

Vectors can be entered in 3 ways: (i) **Author** an expression of the form $[x_1, x_2, \ldots, x_n]$, (ii) use the **Declare, Vector** command which prompts the user for the size and elements of the vector, or (iii) **Author** the in-line command **VECTOR**($f(k), k, n$) to obtain vector elements $f(k), k = 1, 2, \ldots, n$. For example, using the method in item (iii), one can **Author** the expression `VECTOR(k^3,k,4)` and **Simplify** to obtain the vector `[1,8,27,64]`. Other stepping parameters for the indices can be used in the **VECTOR** command. See the *Derive User Manual* for details.

Matrices are entered in similar fashions with vectors acting as rows of the matrix. **Author** `[[1,2,3],[a,b,c],[x,sin x,6!]]` results in the matrix:

$$\begin{bmatrix} 1 & 2 & 3 \\ a & b & c \\ x & \text{SIN}(x) & 6! \end{bmatrix}$$

Similarly, the **Declare, Matrix** command prompts for the dimensions and elements of a matrix. Nested calls to the **VECTOR** command can produce a matrix with function evaluations for values of the elements. The identity matrix of n dimensions is produced with the **IDENTITY_MATRIX(n)** command.

The . (period or dot) is the operator used to perform dot products and matrix multiplication between vectors and matrices. The dimensions of the vectors and matrices must be consistent in order for this operator to execute. The ' key is used to transpose a matrix or vector, and the ^ key is used to produce powers of matrices. The sequence `^-1` attached after the matrix computes the matrix inverse. Descriptions of the most common matrix operations are provided in the following table.

Command	Function
ELEMENT(m,i,j)	Extracts element (i, j) from a vector or matrix m
CROSS(v,w)	Calculates the cross product of 2 vectors $(v \times w)$
DIMENSION(m)	Determines the number of rows in matrix m
OUTER(v,w)	Computes the outer product of vectors v and w
DET(m)	Calculates value of the determinant of matrix m
TRACE(m)	Sums the diagonal elements of matrix m
ROW_REDUCE(m)	Computes the row echelon form of matrix m
CHARPOLY(m)	Produces characteristic polynomial of matrix m
EIGENVALUES(m)	Determines eigenvalues of matrix m by finding the roots of the characteristic polynomial

Matrix manipulation commands and their functions

To solve a system of linear equations, the equations are entered as elements of a vector and the `soLve` command is issued. If there are more variables than equations, the system prompts the user for the variables to solve for in the output. If the system is singular, the solution will either contain arbitrary values (`@1`, `@2`, etc.) or will display the message: `No solutions found`.

Examples of input expressions and results for three linear systems are provided. The input for the first example is to select the `Author` command and enter `[x+y-0.4z=10,y-15z=-0.51,-2x-y+z=1]`.

The solution is found by executing the menu command `soLve`. The resulting solution is shown in expression #2 of the following figure.

The second example is similar to the first, except as shown in expression #4 of the display screen, the output contains `@1` which signifies an arbitrary constant.

The third example contains four variables in the three equations, so Derive queries the user for the variables to be solved for when the `soLve` command is issued. In this case a, b, and c are the solve variables given.

The solution is given in expression #6 in terms of the variable *d*.

```
1:    [x + y - 0.4 z = 10, y - 15 z = -0.51, - 2 x - y + z = 1]

         ┌        77147          157449         2151 ┐
2:       │x = - ───────, y = ───────, z = ────── │
         └        7600            7600          1520 ┘

3:    [6 x + 3 y - z, - 2 x - 11 y + z, 7 x - 14 y]

         ┌                @1          15 @1 ┐
4:       │x = @1, y = ───, z = ───── │
         └                 2             2   ┘

5:    [a + b + c = d, 16 b - d = 13, a - b - c]

         ┌     d         d + 13        7 d - 13 ┐
6:       │a = ──, b = ────────, c = ────────── │
         └     2           16             16    ┘
```

1.9 Complex Variables

Men pass away but their deeds abide.

—last words of Augustin Cauchy [1857]

Derive can perform complex arithmetic and can handle complex variables and complex-valued functions. Variables are declared as complex using the **Declare, Variable** command. The imaginary unit i is entered using **#i** or **Alt-i** and displayed as $\hat{\imath}$. Some of the common functions involving complex numbers are given in the following table with $z = x + iy$.

Command	Function
ABS(z)	$\|x + \hat{\imath}y\|$
SIGN(z)	Point of unit magnitude with same phase angle as z
RE(z)	Real part of z, (x)
IM(z)	Imaginary part of z, (y)
CONJ(z)	Complex conjugate of z
PHASE(z)	Principal phase angle of z

Common functions in Complex Variables

Derive uses a phase angle restricted from $-\pi$ to π radians and by default simplifies roots to the principal branch. Branches can be handled differently by setting modes using the **Manage, Branch** command. See the *Derive User Manual* for an explanation of Derive's use of branches. When Derive is asked to **Factor, Complex** a polynomial, the search is made for the linear complex factors.

1.10 Plotting

There is no royal road to geometry.

— Euclid [300 B.C.]

Derive has the capability to do both 2- and 3-dimensional plotting. Both of these capabilities are available through the **Plot** command of the main menu.

Before the plotting commands are discussed, another important command must be reviewed. Derive has several display modes, so it is important to set the correct mode for your hardware and computational task. The command used to set modes is **Options, Display** (see Section 1.3). It is usually best to set the **Mode** to **Text** when using the algebra screen and to **Graphics** when plotting although the **Graphics** mode can be used for both. See the *Derive User Manual* for more details and useful suggestions. The **F5** key switches to the previous display mode so this key is handy to use to switch back and forth between text and graphics modes. The **Options, Color** command allows for setting of the colors in both the text screen and plotting screen.

The 2-Dimensional **Plot** screen and menu are as shown:

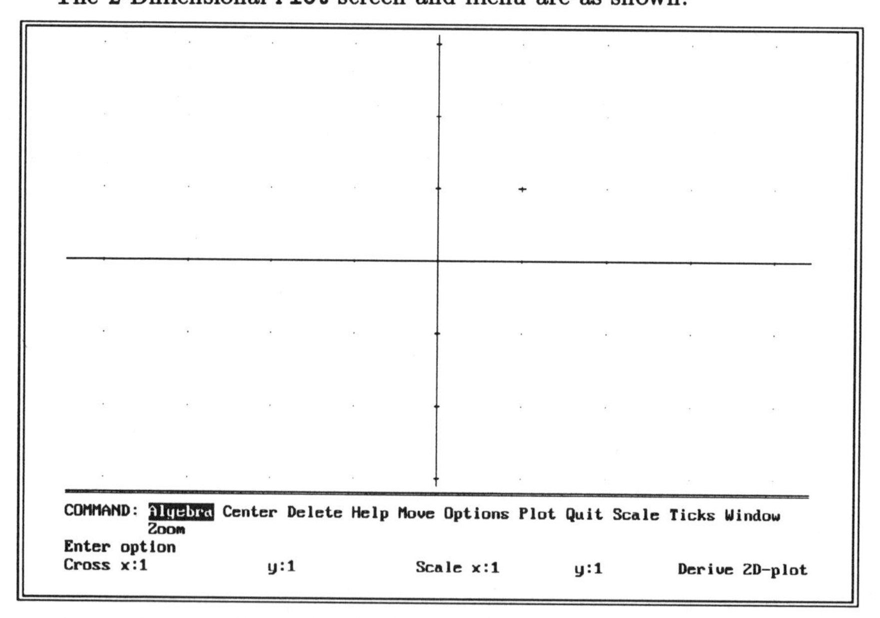

To produce a 2-dimensional plot in rectangular coordinates, the following procedure is used: (i) highlight the expression to plot (it must be in the form $y = f(x)$ or just $f(x)$, but the variable names do not matter), (ii) call the `Plot` menu, and (iii) execute the `Plot` command.

If several functions need to be plotted on the same axes, they should be assembled as components of a vector function. Then all the functions in the vector will be plotted on the same axis at one time. This technique saves considerable time in manipulating the highlight and replotting when the plots are done one at a time.

This produces a plot using the default parameters, which may or may not be sufficient. The plotting parameters are controlled with additional commands in the `Plot` menu. The following table gives a short explanation of these `Plot` commands:

Command	Function
`Algebra`	Sends control back to the algebra screen and menu
`Center`	Positions the center of the plot at the location of the movable cross
`Delete`	Delete functions from the plot list
`Move`	Moves the cross to the desired coordinates
`Options`	Sets the parameters for `Accuracy`, `Color`, `Display`, and `Type`
`Plot`	Draws plots of all the functions in the plot list
`Scale`	Allows user to establish the scale value of each tick mark on the respective axes (see `Zoom` to do this automatically)
`Ticks`	Establishes the distances between tick marks on the axes; can control the aspect ratio of the plot
`Window`	Opens a window; same as the `Window` command in the main menu
`Zoom`	Automatically changes the plot scale by a fixed amount, can zoom `In` or `Out` and in the x, y, or both directions
`Quit`	Stops the execution of the Derive program

2-Dimensional `Plot` menu commands and their functions

There are several helpful keystrokes to know while plotting. The direction arrow keys move the small cross around the plot screen. The coordinates of the cross are displayed in the status line at the bottom of the screen. This cross can be handy in finding approximations to many values of the plotted function. The **F9** and **F10** keys are equivalent to the `Zoom, In` and `Zoom, Out` commands, respectively.

Derive also does 2-dimensional polar plotting by changing the mode to `Polar` using the `Options, Type` command. In polar mode the function is considered to be in the $r = f(\theta)$ or $f(\theta)$ forms even though any variable names can be used.

Similarly, 2-dimensional parametric plotting is available by highlighting a function in the form $[f(t), g(t)]$ with the x (horizontal) direction first and the y (vertical) direction second. Derive plots vectors of functions with more than two components as separate functions in rectangular coordinates. This is a good way to plot three or more functions simultaneously.

Derive's 3-dimensional rectangular surface plots are produced in a similar manner. The highlighted function must be in the form $z = f(x, y)$ or just $f(x, y)$. Once again, the actual variable names are not important. When the `Plot` menu is selected, Derive recognizes the number of variables in the highlighted function and opens the 3-dimensional plot screen. The 3-dimensional plot menu is as shown.

```
COMMAND: Algebra Center Eye Focal Grids Hide Length Options Plot Quit Window
         Zoom
Enter option
Center x:0          y:0          Length x:10       y:10        Derive 3D-plot
```

The commands to control the plot parameters in three dimensions are described in the following table:

Command	Function
Algebra	Sends control back to the algebra screen and menu
Center	Positions the center of the plot at the specified coordinates
Eye	Sets the coordinates of the viewer's eye
Focal	Sets the coordinates of the focal point
Grids	Establishes the number of grid panels in $x-$ and y-directions
Hide	Allows for removal or inclusion of hidden lines
Length	Establishes the lengths of the sides of the box where the plot resides
Options	Allows control over axes display, colors of graphs, and display modes
Window	Opens a window; same as the **Window** command in the main menu
Plot	Produces the plot of the highlighted function
Zoom	Automatically changes the plot scale by a fixed amount, can zoom **In** or **Out**
Quit	Stops the execution of the Derive program

3-Dimensional **Plot** menu commands and their functions

The coordinates of the plot's center and the length of its sides are displayed in the status line at the bottom of the screen.

The following example shows a surface plot of the function $z(x,y)$ defined by

$$z(x,y) = \begin{cases} 9 - x^2 - y^2 & \text{if } 9 - x^2 - y^2 > 0 \\ 0 & \text{otherwise} . \end{cases}$$

This function is converted to a single line expression using the **STEP** function in Derive. It is entered with the **Author** command and by typing

```
(9-x^2-y^2) STEP(9-x^2-y^2)
```

.

The **Plot** menu is selected to open the 3-D plotting window. Then the default plot parameter **Grid** is changed to include 20 grid panels in both the x and y directions and the **Eye** parameters are set to $x = 20$, $y = 15$, and $z = 25$. Finally, the **Plot** command is issued to produce the following surface plot.

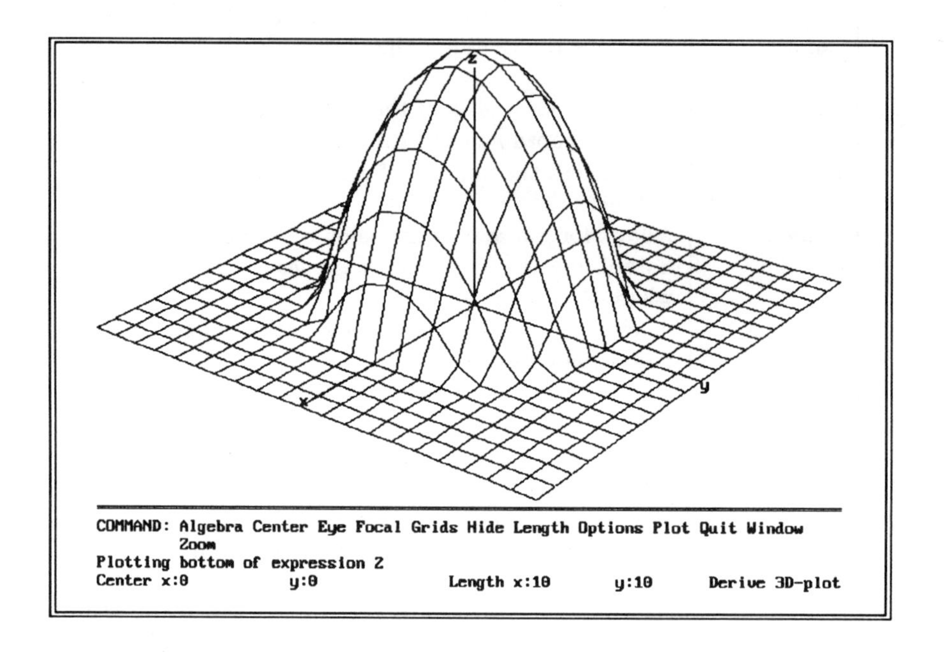
z
x
y
COMMAND: Algebra Center Eye Focal Grids Hide Length Options Plot Quit Window
Zoom
Plotting bottom of expression 2
Center x:0 y:0 Length x:10 y:10 Derive 3D-plot

1.11 Numerical and Functional Approximations

Four Out of Three Jocks Can't Count

—headline in the *Harvard Lampoon* [1986]

When Derive is unable to perform an exact computation or determine the solution symbolically, numerical approximations are viable alternatives. Derive has numerical capabilities in several areas. Some of these capabilities are discussed in this section.

If an antiderivative cannot be determined symbolically for the purpose of evaluating a definite integral, the user can change the **Options, Precision** mode to **Approximate** to allow Derive to perform numerical quadrature. The numerical method is used automatically when Derive is in **Mixed** mode and an exact value cannot be calculated. The approximation method used is adaptive Simpson's quadrature. The method computes the value of the definite integral to an error tolerance of the digital precision established with the **Options, Precision** command. For example, Derive does not provide an answer to $\int_0^{0.2} e^x/(1-x^2)dx$ in **Exact** mode. However, in **Mixed** mode with 6 digits of precision established, the following approximation is produced for this definite integral by using the **Author** command and entering `INT(Alt-e^x/(1-x^2),x,0,0.2)` and then executing **Simplify**.

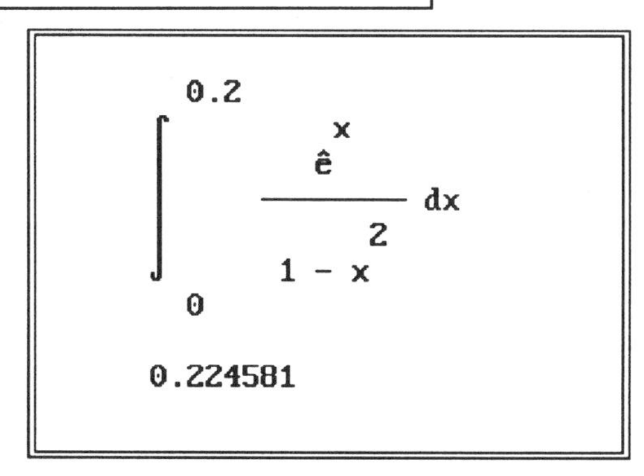

A similar circumstance can occur when solving for roots of an expression of one variable. When the equation is too difficult to solve completely in **Exact** mode, an implicit result is given. If the mode is changed to **Mixed** or **Approximate**, Derive uses the bisection method to find a root within the digital precision established. The user is prompted for the upper and lower bounds of the interval in which to search for the root. For example, if the roots of $e^x - x^4$ are required, in **Exact** or **Mixed** mode no solution is produced. However, in the **Approximate** mode the following result is obtained, if bounds of 0 and 10 are provided after entering the function with the **Author** command and executing the **soLve** command to get the root as shown.

```
 x    4
ê  - x

x = 8.61314
```

The **Calculus, Taylor** command finds a Taylor polynomial approximation to an expression. This command can be executed through the menu or in-line as TAYLOR($f(x), x, x_0, n$), where $f(x)$ is expanded in an nth degree polynomial about x_0. Taylor approximations can be handy for integration when direct symbolic integration is not possible.

It is usually best to use Derive in the **Exact** precision mode and convert to the **Approximate** mode when numerical approximated is desired. When decimal simplification of irrational or rational numbers is needed, they can be approximated by using the **approX** command.

Derive is not designed to perform sophisticated manipulation of large data sets. However, it has several functions that perform data analysis when the data is placed into a 2-column matrix. One such function is FIT(m) which provides a least squares approximation to the data in the functional form provided. See the *Derive User Manual* for a description of this command.

Two operations (PICARD and TAY_ODE1) in the utility file ODE1.MTH produce approximate solutions to differential equations. See Section 1.2 and Examples 2.18 and 2.19 for information on these operations.

1.12 Utility Files for Differential Equations

I do not think that one could acquire any solid nature in physics without geometry, and the best of geometry is analysis.

—Debeaune to Mersenne [1639]

Utility files are pre-established files of Derive expressions saved for use at a later time. The Derive disk comes with several utility files. Some of the utility files containing calculus operations were discussed in Section 1.7. Users can build and save their own utility files using the **Transfer, Save** command. There are two utility files in Derive Version 1.60 and later that contain functions to help solve differential equations. The two files are ODE1.MTH for first-order equations and ODE2.MTH for second-order equations. Several of the functions in these files will be used in the Examples in Chapter 2. These two utility files contain short comments describing the functions' purposes and parameters. The proper type and form of the equation must be determined before the functions can be used. However, these built-in functions make Derive a powerful and efficient tool for solving differential equations. The functions in these two files and their purposes are given in the following tables along with references to the examples and exercises of Chapters 2 and 3 where they are used. Further details are provided when the functions are used in examples of Chapter 2.

Function Name	Purpose
`SEPARABLE(p,q,x,y,x0,y0)`	Solves the separable differential equation, $y' = p(x)q(x)$, $y(x0) = y0$ (Ex. 2.1)
`EXACT_IF_0(p,q,x,y)`	Checks if equation $p + qy' = 0$ is exact
`EXACT(p,q,x,y,x0,y0)`	Solves the exact equation (above), with $y(x0) = y0$ (Ex. 2.2)
`USE_INTEG_FCTR (m,p,q,x,y,x0,y0)`	Solves equation with known integrating factor m
`LINEAR1(p,q,x,y,x0,y0)`	Solves linear equation $y' + py = q$, with $y(x0) = y0$ (Ex. 3.20)
`BERNOULLI (p,q,k,x,y,x0,y0)`	Solves the Bernoulli equation $y' + py = qy^k$, with $y(x0) = y0$ (Ex. 3.20)
`HOMOGENEOUS_IF_FREE_OF_X (r,x,y)`	If this returns a 0, the equation $y' = r(x, y)$ is homogeneous
`HOMOGENEOUS(r,x,y,x0,y0)`	Solves the homogeneous equation (shown above)

Elementary functions for solving first-order equations in the Utility File ODE1.MTH

Function Name	**Purpose/Form of Equation to be Solved**
`FUN_LIN_CCF (q,a,b,c,x,y,x0,y0)`	Solves $y' = q(x,y)(ax+by+k)$, with $y(x0) = y0$
`LIN_FRAC (r,a,b,c,s,t,d,x,y,x0,y0)`	Solves the equation: $y' = r(x,y)(ax+by+c)/(sx+ty+d)$, $y(x0) = y0, sa - eb \neq 0, cd \neq 0$
`INTEG_FCTR_FREE_OF_X (r,p,q,x,y,x0,y0)`	Solves $p(x,y)+q(x,y)y' = 0$, when the integration test is free of x and gives r
`INTEG_FCTR_FREE_OF_Y (r,p,q,x,y,x0,y0)`	Solves $p(x,y)+q(x,y)y' = 0$, when the Integration test is free of y (similar to above)
`GEN_HOM(r,k,x,y,x0,y0)`	Solves general homogeneous equation $y' = r(x,y) = h(yx^k)y/x$
`ALMOST_LIN (r,h,p,q,x,y,x0,y0)`	Solves $r(x,y)y' + p(x)h(y) = q(x)$, with $y(x0) = y0$
`CLAIRAUT`	Helps solve the Clairaut equation (arguments not given)
`TAY_ODE1(r,x,y,x0,y0)`	Finds 4th degree Taylor-series solution to $y' = r(x,y), y(x0) = y0$ (Ex. 2.17)
`PICARD(r,y_prev,x,y,x0,y0)`	Given approximate solution y_{prev} to $y' = r(x,y)$, with $y(x0) = y0$, finds an improved iterate

Functions for the solution of certain first-order equations using the advanced methods found in Utility File ODE1.MTH

In the functions of ODE2, linear equations in the form $y'' + p(x)y' + q(x)y = r(x)$ are called *reduced* if $r = 0$ or *complete* if $r \neq 0$. *Homogeneous* is reserved for equations invariant under $x = tx$ and $y = ty$.

Function Name	Purpose/Form of Equation to be Solved
LIN2_RED_CCF_DISC(p,q)	Determines the value of the discriminant $p^2 - 4q$ for $y'' + py' + qy = 0$
LIN2_RED_CCF_POS(p,q,x)	Solves $y'' + py' + qy = 0$, if $p^2 - 4q > 0$
LIN2_RED_CCF_NEG(p,q,x)	Solves $y'' + py' + qy = 0$, if $p^2 - 4q < 0$
LIN2_RED_CCF_0(p,x)	Solves $y'' + py' + qy = 0$, if $p^2 - 4q = 0$
LIN2_COMPLETE(u,v,r,x)	Solves for the particular solution of $y'' + py' + qy = r(x)$ given the complementary solutions to the reduced equation
SECOND_RED_PARTIC(p,u,x)	Given one solution of $y'' + py' + qy = 0$, finds second linearly independent solution
AUTONOMOUS_CONSERVATIVE (q,x,y,x0,y0,v0)	Solves $y'' = q(y)$, with $y(x_0) = y_0$ and $y'(x_0) = v_0$
IMPOSE_BV2 (x,y,x1,y1,x2,y2)	Given $y(x, c_1, c_2), y(x1) = y1$, $y(x2) = y2$, finds c_1 and c_2
IMPOSE_IC2(x,y,x0,y0,v0)	Given $y(x, c_1, c_2), y(x0) = y0$, $y'(x0) = v0$, finds c_1 and c_2

Functions for the solution of certain second-order equations using the basic methods found in Utility File ODE2.MTH

Function Name	Purpose/Form of Equation to be Solved
`LIOUSVILLE(p,q,x,y)`	Finds the general solution of $y'' + p(x)y' + q(y)(y')^2 = 0$
`EXACT2(f,g,x,y,v)`	Finds one solution to the exact equation $f(x,y,y')y'' + g(x,y,y') = 0$, where $v = y'$
`EULER(p,q,b)`	Used to convert an Euler equation, $y'' + py'/(ax+b) + qy/(ax+b)^2 = 0$, to a constant coefficient equation; solve the new equation and transform it by $t \to \ln(ax+b)$
`EQUIDIMENSIONAL_Y`, `EQUIDIMENSIONAL_X`	These two commands are used to solve $y'' = r(x,y,y')$, when it is equidimensional. (check the file for explanation of arguments)
`HOMOGENEOUS2(r,x,y,v)`	Used to solve $y'' = r(x,y,y')$, when it is homogeneous with $v = y'$

Some of the functions for the solution of certain second-order equations using the advanced methods found in Utility File ODE2.MTH. There are other functions available in the file.

In order to use any of these functions, the appropriate file (ODE1 or ODE2) must be loaded into the work area using the **`Transfer, Load`** command. Users can make their own utility files and save them for future use using the **`Transfer, Save`** command. To save loading time, memory space, and screen clutter, it is probably a good idea to split utility files ODE1 and ODE2 into elementary and advanced parts. First courses in differential equations generally use only the elementary solution techniques. Just such an elementary utility file is outlined in Section 1.14.

Derive also has a utility file, RECUREQN.MTH, for solving first and second order difference equations. See Sections 2.15 and 3.12 for more details concerning this file.

1.13 Limitations

What we know is very slight. What we don't know is immense.

—Laplace's last words [1827]

All software packages have limitations, and Derive is no exception. It is usually best to be warned of those limitations before discovering them first-hand at a critical step in solving a problem.

Derive's symbolic manipulation is its strength, but it doesn't always find closed form solutions that exist. Sometimes, it doesn't simplify as much as you would like. So don't think Derive is going to do all the manipulations you will need to solve a problem.

Probably Derive's most obvious restriction is its lack of any programming capability. Derive's commands are executed one at a time with user interface needed after each step. However, through the use of nested functional calls, several operations can be executed at once. But, because of this limitation, there is no way to do iteration, recursion, or branching. Inconveniences related to this situation are that no subscripted variables are possible and that Derive has no way to establish integer data types (so, for example, $\sin n\pi$ is not recognized as equalling 0).

Because the software is totally menu-driven, the user interface is crucial. Yet Derive does not support the use of a mouse to facilitate this interface. This sometimes slows cursor movement, highlighting, and selection of menu options.

Derive's work area is designed for executable commands, not text. Therefore, there is very little text formatting allowed in the work area. This restricts the quality of the screen image and printed output. However, in **Graphics** mode, Derive's display of mathematical symbols is impressive. It is sometimes annoying that Derive does not allow expressions to wrap-around on the next line of the screen. Using the direction arrows to read the rest of the expression can be tedious. It would help if the wrap-around feature was an optional capability.

There are several numerical tools that, if present, could help in problem solving. Among the most missed for the tasks at hand are a numerical differential equation solver, a numerical solution technique for systems of nonlinear equations, and a direct method to compute eigenvectors. In addition, a way to perform contour integrals would be helpful.

Similarly, the limitations in Derive's plotting tools are the *lack* of
i) direct entry of domain bounds for a two-dimensional, rectangular plot, ii) an automatic scaling option, iii) the ability to plot data points, iv) the

ability to handle implicit functions, v) the ability to plot level curves for a surface, and vi) the ability to plot a direction field. It is very difficult to get plots in specified bounds. On the other hand, the resolution of plots is first-rate.

The only task that sometimes seems slow in Derive is plotting. Despite the user's ability to change accuracy parameters to speed up plotting, it seems that some plots take too long, especially when several are needed. Although the ability to plot a vector of expressions with one plot command is very helpful, users should set aside plenty of time when performing multiple plots of complicated functions. See Example 2.13 for plotting a vector of expressions.

While some of these limitations are inherent to Derive's design, other limitations most likely will be eliminated in future versions of Derive.

1.14 Recommendations

...it is unworthy of excellent men to lose hours like slaves in the labor of computation.

—Gottfried Wilhelm Leibniz

There are several things that can be done to insure efficient and proper use of the tremendous capabilities of Derive and to tailor the software to the specific needs of the user. First, insure the system state you use most often is saved into the DERIVE.INI file and, therefore, is in effect when the program is started. This is accomplished by putting the computer in the state you want and then saving that state using the **Transfer, sTate, Save** command. If you use the plotting features often, insure this initial state is in the appropriate **Graphics** mode. This will eliminate the need to change states very often in order to get the capabilities you need. See Sections 1.2 and 1.5 and the *Derive User Manual* for more information on this feature.

Next, build and save utility files that contain the functions that you use in a manner that is efficient for your use. For instance, select the functions you use most often in the utility files ODE1.MTH and ODE2.MTH and design your own utility file for solving differential equations. You may want to add other functions like **LAPLACE** and **FOURIER** from the INTEGRAL.MTH utility file to your file, if you frequently use those commands. The following table lists some functions, along with the Derive utility file they are found in, that may produce a nice utility file for an elementary differential equations course.

Function Name	Utility File
SEPARABLE	ODE1
EXACT_IF_0	ODE1
EXACT	ODE1
USE_INTEG_FCTR	ODE1
LINEAR1	ODE1
TAY_ODE1	ODE1
PICARD	ODE1
LIN2_RED_CCF_DISC	ODE2
LIN2_RED_CCF_POS	ODE2
LIN2_RED_CCF_NEG	ODE2
LIN2_RED_CCF_0	ODE2
LIN2_COMPLETE	ODE2
SECOND_RED_PARTIC	ODE2
IMPOSE_BV2	ODE2
IMPOSE_IC2	ODE2
LAPLACE	INTEGRAL
FOURIER	INTEGRAL

Suggested functions for a utility file for an elementary differential equations course

If you frequently repeat a sequence of commands or use a command in a special way, try to design your own commands and add them to appropriate utility files.

Finally, spend some time learning the keystrokes that help manipulate expressions and reduce retyping expressions. These keystrokes are discussed in Section 1.3. Derive's menu system makes it user-friendly; however, the understanding and use of a few of the special keystrokes can turn you into a more efficient user of this powerful software tool.

Chapter 2

Examples

The heart of mathematics consists of concrete examples and concrete problems.

—Paul R. Halmos [1970]

These examples have been worked out to show the power and versatility of Derive as a problem-solving tool. In carefully reading these problems and solutions and working along with Derive, the reader should get a better feel for the subject of differential equations and of mathematical problem solving in general. Some of these examples also show the limitations of the computer, and help the reader to know when to use Derive and the computer and when not to.

Many of these examples involve models of realistic applications, while others are posed in a mathematical context. The problems are similar to those typically found in undergraduate applied differential equations textbooks.

Example 2.1: Separable Differential Equation

Eventually there will be, I hope, some people who will find it profitable to decipher this mess.

—Evariste Galois [1832]

Subject: Solving the Logistics Equation, a Separable Differential Equation, using Utility File ODE1

Problem: Find the solution to the modified logistics equation

$$\frac{dP}{dt} = P(a - bP)(1 - cP^{-1})$$

with $a, b, c > 0$ and $P(0) = 1000$.

Solution: This is a separable differential equation, therefore the operation `Separable` in file ODE1 can be used to solve the equation. The utility file ODE1 is loaded using the **`Transfer, Load`** command and giving `ODE1` for the file name. This procedure takes a while since over 60 expressions are read into the work area.

If the differential equation is in the form $y' = p(x)q(x)$ with $y(x_0) = y_0$, then the form of the operation to solve the equation is:

$$\texttt{SEPARABLE}(p(x), q(y), x, y, x_0, y_0).$$

Before this operator is entered, the given conditions on the parameters a, b, and c must be established. This is done with the **`Declare, Variable`** command making a, b, and c all **`Positive`**. This has to be done one variable at a time. The keystrokes **d v** result in the following menu display:

```
DECLARE variable:
Enter variable or type DEFAULT
                                   Free:100%            Derive Algebra
```

Then the sequence of screen displays showing the input to declare the variable a positive are as follows:

```
DECLARE variable: a

Enter variable or type DEFAULT
                                    Free:100%          Derive Algebra
```

```
DECLARE: Domain: Positive Nonnegative Real Complex Interval

Select domain of a
                                    Free:100%          Derive Algebra
```

Now, to solve this equation, execute the **Author** command and enter:

`SEPARABLE(1, P(a-bP)(1-c/P),t,P,0,1000)`.

The command line and working area shows this as:

```
AUTHOR expression: separable(1,p(a-bp)(1-c/p),t,p,0,1000)_

Enter expression
                                    Free:100%          Derive Algebra
```

```
                                        c
68:  SEPARABLE [1, p (a - b p) [1 - ———], t, p, 0, 1000]
                                        p

COMMAND: Author Build Calculus Declare Expand Factor Help Jump soLve Manage
         Options Plot Quit Remove Simplify Transfer moVe Window approX
Enter option
User                  D:ODE1.MTH          Free:84%  Insert      Derive Algebra
```

Simplify this expression to obtain the implicit solution:

$$69: \quad \frac{\text{LN}\left[\frac{c - p}{a - b\,p}\right]}{a - b\,c} - \frac{\text{LN}\left[\frac{c - 1000}{a - 1000\,b}\right]}{a - b\,c} = t$$

In order to obtain an explicit form for $P(t)$, ask Derive to **soLve** for P. The result is

$$70: \quad \mathrm{LN}\left[\frac{p - c}{b\,p - a}\right] = \mathrm{LN}\left[\frac{c - 1000}{a - 1000\,b}\right] + t\,(a - b\,c)$$

This indicates Derive cannot solve explicitly for P. *Don't give up.* Help Derive by making it do what you would try. Exponentiate both sides of the equation. This can be done without retyping the terms in the equation by making use of the keystroking hints in Section 1.3. The procedure and keystrokes to this are (i) highlight the left side of the equation using the ← key, (ii) `Author` and enter **Alt-e^F4**, (iii) highlight the right side of the equation using the → key, and (iv) enter **= Alt-e^F4**. The working expression should be

$$71: \quad \hat{e}^{\mathrm{LN}\,((p - c)\,/\,(b\,p - a))} = \hat{e}^{\mathrm{LN}\,((c - 1000)\,/\,(a - 1000\,b)) + t\,(a - b\,c)}$$

Now `soLve` for P to get

$$73: \quad p = \frac{a\,(c - 1000)\,\hat{e}^{a\,t} - c\,(a - 1000\,b)\,\hat{e}^{b\,c\,t}}{b\,(c - 1000)\,\hat{e}^{a\,t} + (1000\,b - a)\,\hat{e}^{b\,c\,t}}$$

$$74: \quad p = \frac{1}{0}$$

It is obvious that the first equation is the desired solution.

To check the initial condition, `Manage, Substitute` into the expression for the solution, the value 0 for the variable t. Just press **Enter** when queried for replacement for the other unknowns. The result is

$$75: \quad p = \frac{a\,(c - 1000)\,\hat{e}^{a\,0} - c\,(a - 1000\,b)\,\hat{e}^{b\,c\,0}}{b\,(c - 1000)\,\hat{e}^{a\,0} + (1000\,b - a)\,\hat{e}^{b\,c\,0}}$$

Now, `Simplify` this expression to achieve the given initial value of 1000.

To check the solution, **Author** and enter the operator shown in the following input line.

```
AUTHOR expression: dif(p,t)-p(a-bp)(1-c/p)

Enter expression
Simp(75)              D:ODE1.MTH            Free:64%           Derive Algebra
```

Then, **Manage, Substitute** the expression for the solution P (for this example, this is the right-hand side of expression #73) into the operator using the highlight and **F3** key. **Simplify** to get the following result:

```
             2              2  2 a t                                  a t + b c t
        b (a  (c - 1000)  ê          - 2 a c (c - 1000) (a - 1000 b) ê            +
80:  - ---------------------------------------------------------------------------
                                              a t                    b c t 2
                                 (b (c - 1000) ê     + (1000 b - a) ê       )
```

which unfortunately doesn't verify the result. We expect this to be equal to 0. Let's try substitution of specific values ($a = 0.1, b = 0.2$, and $c = 0.4$ for example) using **Manage, Substitute**. Then execute **Simplify** to obtain the desired result of 0. This kind of messy result doesn't happen often with Derive. However, there must be something in the symbolic expression that Derive does not simplify enough in order to notice that everything cancels out to leave 0. Once numerical values are substituted Derive recognizes the cancellation. Later, in this manual you'll see Derive handle cases similar to this with no problem.

Manage, Substitute these same values for a, b, and c into the solution equation and **Simplify**. The output is

```
                             0.1 t                              0.2 0.4 t
           0.1 (0.4 - 1000) ê       - 0.4 (0.1 - 1000 0.2) ê
83:  p = ------------------------------------------------------------------
                                  0.1 t                      0.2 0.4 t
             0.2 (0.4 - 1000) ê       + (1000 0.2 - 0.1) ê

                      t/50
           2 (2499 ê       - 1999)
84:  p = -------------------------
                      t/50
             9996 ê       - 9995
```

Plot, Plot this solution with the default scale to show the solution graphically with these parameter values as:

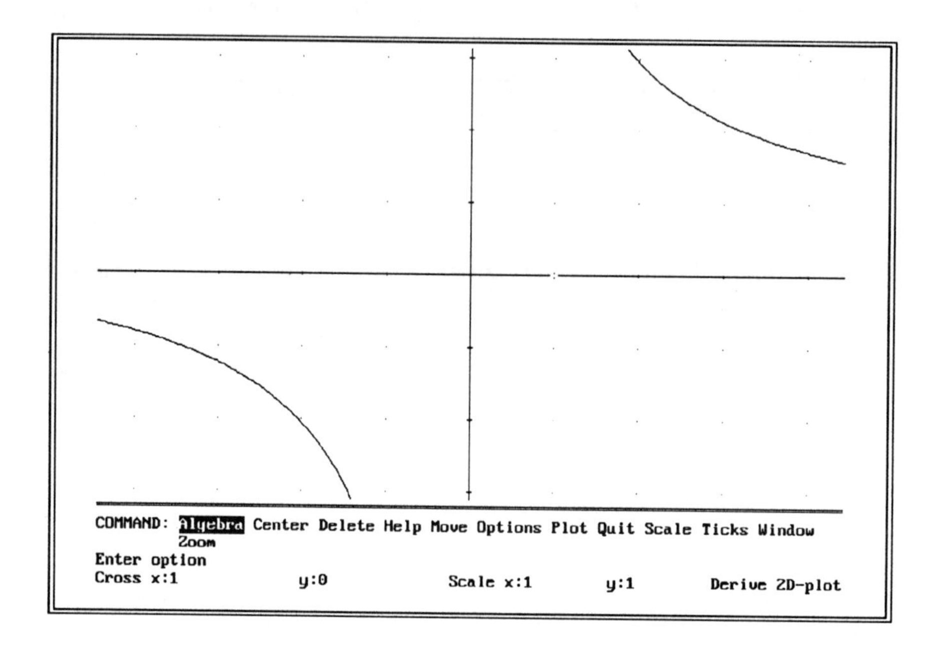

Unfortunately, this plot does not show much of the overall behavior of the solution. This is because the default plotting window shows approximately $-4.5 < x < 4.5$ and $-3 < y < 3$ and the interesting behavior for this function lies outside of that window. We will be changing the window size and using more of the plotting capabilities by changing the plotting parameters in upcoming example problems.

Example 2.2: Exact Differential Equation

If enthusiasm continues, it won't be long before every campus is "infected" by technology ...

—article in *UME Trends* [1989]

Subject: Solving an Exact Differential Equation using Utility File ODE1

Problem: Find the particular solution of

$$(y^2 \cos x - 3x^2 y - e^x) + (2y \sin x - x^3 + \ln y)y' = 0, \quad \text{with } y(0) = 1.$$

Solution: This is a first-order differential equation which may be exact, therefore the operation **EXACT_IF_0** in file ODE1 can be used to check for exactness. The utility file ODE1 is loaded using the `Transfer, Load` command and giving `ODE1` for the file name.

The differential equation is already in the proper form of $p(x,y) + q(x,y)y' = 0$, therefore the form of the exactness check is

EXACT_IF_0$(p(x,y), q(x,y), x, y)$.

The input command for the exactness check for this problem is

```
EXACT_IF_0(y^2cosx-3x^2y-Alt-e^x, 2y sinx-x^3+lny,x,y)
```

This displays in the work area as

```
                     2            2     x                      3
19:  EXACT_IF_0 (y  COS (x) - 3 x  y - ê , 2 y SIN (x) - x  + LN (y), x, y)
```

Then `Simplify` to obtain 0, which indicates the equation is exact.

In order to solve an exact equation, the Command **EXACT** is used. The argument of the command are the same as those of **EXACT_IF_0** along with the initial condition of the equation. For this problem, the command is entered as:

```
AUTHOR expression: exact(y^2COS(x)-3x^2y-ê^x,2ySIN(x)-x^3+LN(y),x,y,0,1)_

Enter expression
User                 D:ODE1.MTH          Free:84%  Insert        Derive Algebra
```

Most of this expression can be entered without retyping by using the highlight and **F3** key to obtain parts of the previous expression (see Section 1.3). **Simplify** this expression to obtain the solution. The command and its result are shown below.

```
                 2             2     x                 3
21:  EXACT (y  COS (x) - 3 x  y - ê , 2 y SIN (x) - x  + LN (y), x, y, 0, 1)

        x              2         3
22:  - ê  + y LN (y) + y  SIN (x) - x  y - y + 2 = 0
```

We now try to solve explicitly for y using the **soLve** command. This is done by keystroking **1 Enter y Enter**. However, the result shows there is a problem. The unchanged expression implies that Derive can't help. This is unfortunate since Derive does not plot implicit functions. However, don't despair, since we do have the solution to the differential equation because of Derive's help.

Example 2.3: Homogeneous Differential Equation

> ***The study of Euler's works will remain the best school for the different fields of mathematics and nothing can replace it.***
>
> —Carl Friedrich Gauss

Subject: Solving a First-Order Homogeneous Differential Equation using Utility File ODE1

Problem: Find the particular solution of

$$y' = \frac{x^2+y^2}{xy}; \quad \text{with } y(1) = -2.$$

Solution: This is not exact or separable (for an exactness check see Example 2.2), but it may be homogeneous. Remember, homogeneous means the equation is invariant under the transformations $x = tx$ and $y = ty$. To check to see if the equation is homogeneous, the operation **HOMOGENEOUS_IF_FREE_OF_X** in file ODE1 can be used. ODE1 is loaded using the **Transfer, Load** command and giving `ODE1` for the file name.

The differential equation is already in the proper form of $y' = r(x,y)$, therefore the form of the check for homogeneity is

HOMOGENEOUS_IF_FREE_OF_X$(r(x,y), x, y)$.

For this problem, **Author**

`HOMOGENEOUS_IF_FREE_OF_X((x^2+y^2)/xy,x,y)`.

and check for homogeneity using the command **Simplify** to produce:

```
                                   ┌  2    2        ┐
                                   │ x  + y         │
22:  HOMOGENEOUS_IF_FREE_OF_X      │ ────────, x, y │
                                   └   x y          ┘

       2
      y  + 1
23:  ──────
        y
```

This expression is free of the variable x, so the equation is homogeneous.

In order to solve this homogeneous equation, the expression $r(x, y)$ with its initial condition is entered using **Author** into the operator **HOMOGENEOUS** as shown in this display of the input line.

```
AUTHOR expression: homogeneous((x^2 + y^2) / (x y),x,y,1,-2)_

Enter expression
User                              Free:100%              Derive Algebra
```

This expression can be entered without retyping using the highlight and **F3** key (See Section 1.3). The working area then shows

```
                    ┌  2    2                  ┐
                    │ x  + y                   │
24:  HOMOGENEOUS    │ ───────, x, y, 1, -2     │
                    └   x y                    ┘
```

Simplify this to obtain the solution

```
          2
         y
25:    ─────  - 2 = LN (x)
          2
       2 x
```

This equation can be solved explicitly for y using the **soLve** command and entering y as the solve variable to obtain

```
26:  y = - √2 √(LN (x) + 2) |x|

27:  y = √2 √(LN (x) + 2) |x|
```

The negative expression is the solution consistent with the initial condition. This equation is easily plotted using the **Plot** command. Once in the plot screen, use the **Scale** menu command to set the **x scale** to 2 and the **y scale** to 10. Remember from the table in Section 1.10 that this command sets the distance between each tick mark on each axis to the scale value entered. Then use the **Move** command to move the cross to $x = 3$ and $y = 0$. Execute the **Center** command to move the plot region so it is centered at the point (3,0). Finally, issue the **Plot** command to plot the function.

Insure the proper function is highlighted back in the algebra window. The resulting plot is shown.

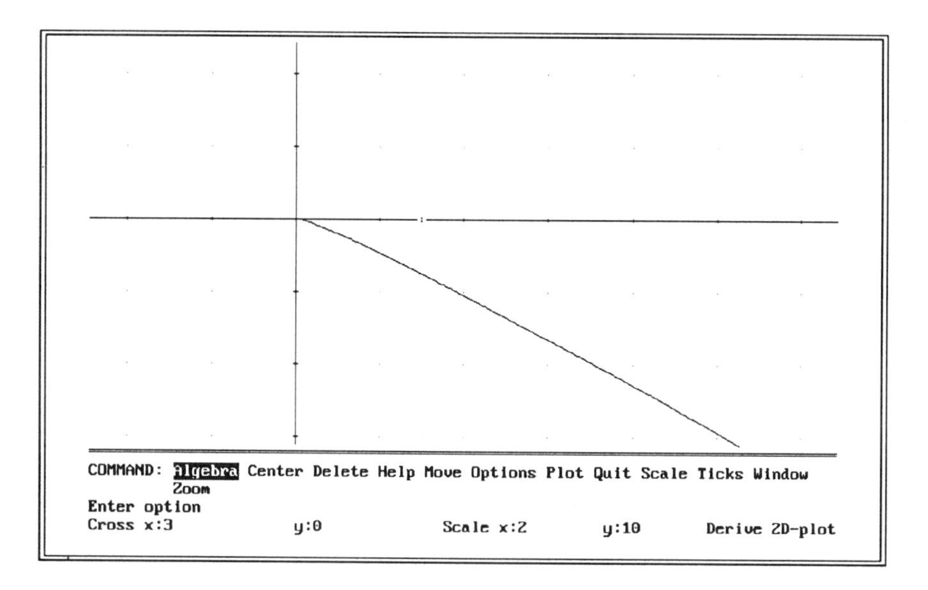

Example 2.4: Variation of Parameters

Numerical precision is the very soul of science.

—Sir D'Arcy Wentworth Thompson [1917]

Subject: Solving a Second-Order, Constant Coefficient, Nonhomogeneous Equation by Variation of Parameters using Utility File ODE2

Problem: Find the general solution to

$$y'' - 2y' + y = 12e^x - 4xe^x$$

with $y(0) = 1$ and $y'(0) = 2$.

Solution:

The operations for solving this kind of second-order, nonhomogeneous differential equation with constant coefficients are found in the utility file ODE2. Load this file with the **Transfer, Load** command. The first step in solving this equation is to find the complementary solution, y_c, of the homogeneous equation of the form $y'' + py' + qy = 0$. This is done by first determining the value of the discriminant $p^2 - 4q$ using the operation **LIN2_RED_CCF_DISC**(p, q). The display of this operation in the work area and its result after executing **Simplify** for this problem are shown.

```
AUTHOR expression: lin2_red_ccf_disc(-2,1)

Enter expression
User                                Free:100%               Derive Algebra
```

```
49:  LIN2_RED_CCF_DISC (-2, 1)

50:  0
```

Since the value of the discriminant from this command is 0, the operation **LIN2_RED_CCF_0**(p, x) is used to solve the equation in homogeneous form. The operators **LIN2_RED_CCF_POS**(p, q, x) and **LIN2_RED_CCF_NEG**(p, q, x) are used to solve equations with positive and negative discriminants, respectively. The operation for this problem and the resulting solution, y_c, obtained by executing **Simplify** are shown.

```
AUTHOR expression: lin2_red_ccf_0(-2,x)

Enter expression
Simp(49)                              Free:100%          Derive Algebra
```

```
51:  LIN2_RED_CCF_0 (-2, x)

                    x
52:  (@2 x + @1) ê
```

The symbols @1 and @2 are the arbitrary constants in the complementary solution. Therefore, the fundamental set of solutions are e^x and xe^x.

The particular solution, y_p, is determined using the method of variation of parameters with the operation **LIN2_COMPLETE**$(u(x), v(x), r(x), x)$, where $u(x)$ and $v(x)$ are the fundamental solutions and $r(x)$ is the forcing function. For this problem the Derive command, its display, and the resulting y_p are shown.

```
AUTHOR expression: lin2_complete(ê^x,xê^x,12ê^x-4xê^x,x)

Enter expression
Simp(51)                              Free:100%          Derive Algebra
```

```
                         x      x       x         x
53:  LIN2_COMPLETE (ê , x ê , 12 ê  - 4 x ê , x)

             2          x
          2 x  (x - 9) ê
54:  - ─────────────────
               3
```

Now, the general solution, y_g, is formed by summing y_c and y_p. This solution can be checked by applying the differential operator. The operator for this equation is entered using **Author** as shown in the following input line.

```
AUTHOR expression: dif(y,x,2)-2dif(y,x)+y

Enter expression
User                                  Free:100%          Derive Algebra
```

The general solution, y_g, is formed by summing the two expressions for y_p and y_c using the highlight and **F4** key. This new expression can be entered into the functional operator using the **F3** key to avoid retyping the long expression. The expression for y_g in Derive format is

```
                                      2         x
                        x          2 x  (x - 9) ê
55:   (@2 x + @1) ê   + - ─────────────────────
                                        3
```

Manage, Substitute the solution into this differential operator (the **F3** key can help). The working area display of this command (truncated on the right because of its length) is shown.

```
                                              2           x
      ┌d ┐2 ┌                    x         2 x  (x - 9) ê  ┐       d  ┌                    x       2
57:   │─ │  │(@2 x + @1) ê   + - ─────────────────────      │ - 2 ─  │(@2 x + @1) ê   + - ─
      └dx┘  └                                 3            ┘       dx └
```

Execute **Simplify** to show that the expression satisfies the equation by seeing that the evaluation of the differential operator equals the right-hand side of the equation.

```
                         x
58:   - 4 (x - 3) ê
```

The two initial conditions are used to solve for the two arbitrary constants with the command **IMPOSE_IC2**(x, y, x_0, y_0, v_0). For this problem the input, its display, and its result upon issuing **Simplify** are shown.

```
AUTHOR expression: impose_ic2(x,(@2 x + @1) ê^x - 2 x^2 (x - 9) ê^x / 3,0,1,2)_

Enter expression
User                                    Free:100%            Derive Algebra
```

```
                                           2            x
                                  x     2 x  (x - 9) ê
59:  IMPOSE_IC2 [x, (@2 x + @1) ê  -  ─────────────────, 0, 1, 2]
                                              3

60:  [ @1 = 1  @2 = 1 ]
```

The Derive display of the final solution is shown.

```
               3        2                x
           (2 x  - 18 x  - 3 x - 3) ê
63:  -  ─────────────────────────────────
                       3
```

The plot of this solution with the default plotting parameters using the **`Plot, Plot`** command produces

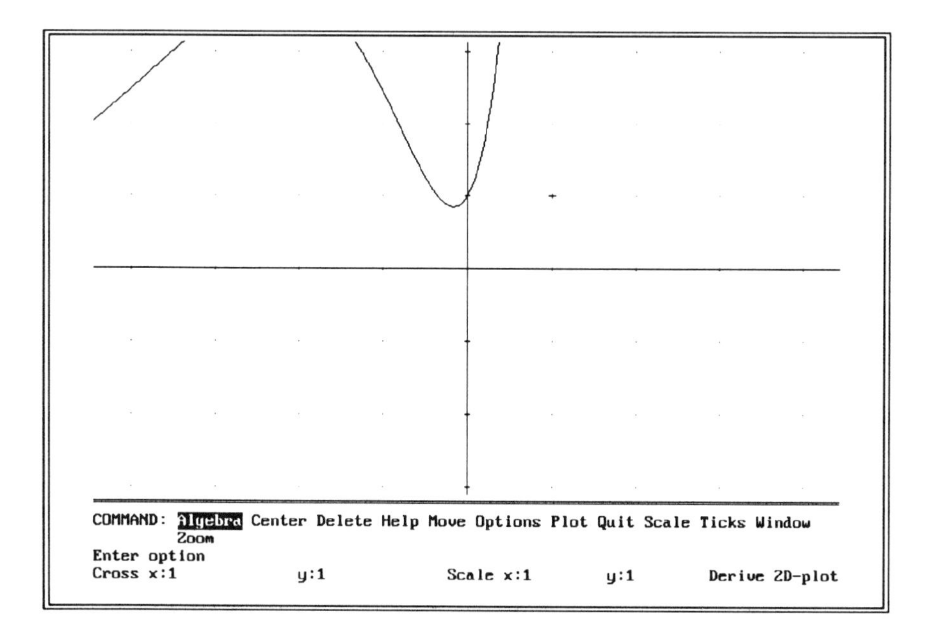

The scale needs to be adjusted to see the global nature of the solution. We try setting the **Scale** in the x-direction to 1 and the y-direction to 10000, using the **Move** command to move the cross to (2500), and executing the **Center** command to better center the interesting region of the plot on the screen. These plotting parameters produce a better graph that shows the exponential growth of the solution. This plot is shown below:

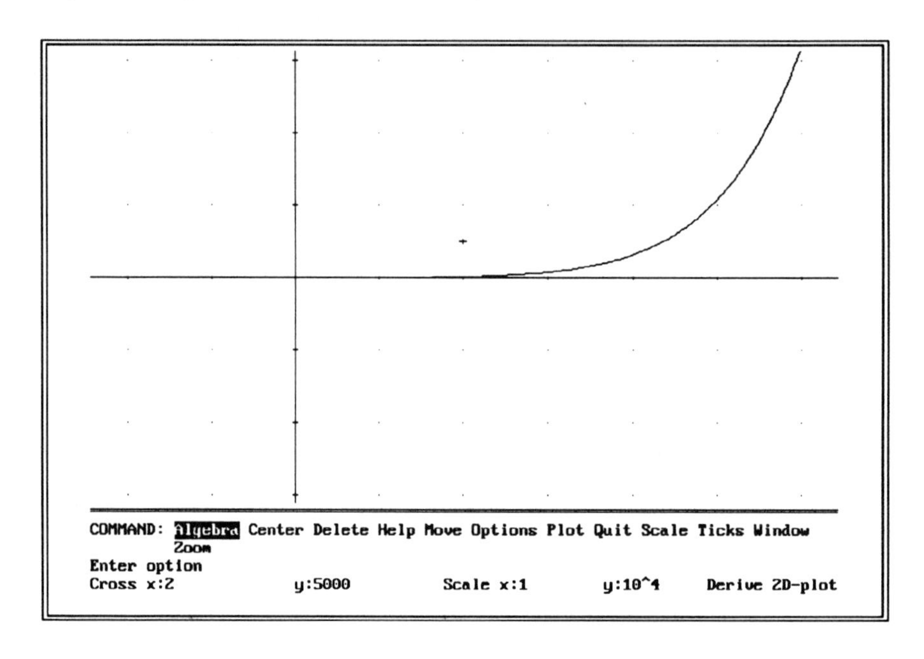

Example 2.5: Undetermined Coefficients

If you do not rest on the good foundation of nature, you will labor with little honor and less profit.

— Leonardo da Vinci

Subject: Solving a Complete, Third-Order, Constant Coefficient, Initial Value Problem with Undetermined Coefficients

Problem: Find the solution to the following initial value problem using undetermined coefficients:

$$L[y] = y''' + y'' + y' + y = e^{-x} + kx,$$

$$y(1) = 1, \; y'(1) = 1/4, \; y''(1) = -1/2, \quad -8 \le k \le 8.$$

Analyze the solution to determine the value of k that maximizes the solution for $x > 1$ and plot this solution.

Solution: The cubic auxiliary equation for this third-order differential equation is $m^3 + m^2 + m + 1 = 0$. Its roots are obtained by inputting this equation using the **Author** command and solving the equation using the **soLve** command. The output from these steps is

```
       3    2
1:    m  + m  + m + 1 = 0

2:    m = -1

3:    m = -î

4:    m = î
```

Therefore, the complementary solution of the homogeneous equation is $y_c = c_1 e^{-x} + c_2 \cos x + c_3 \sin x$.

By using the forms of the parts of the forcing function $e^{-x} + kx$ and this complementary solution, the form of a particular solution can be written by using the method of undetermined coefficients as $Axe^{-x} + Bx + C$. In order to determine these coefficients A, B, and C, the left-hand side $L[y_p]$ is evaluated and equated to the forcing function. To do this with Derive, the operator $L[y]$ is defined using the in-line command for differentiation,

DIF (See Section 1.3). This entry is done with the **Author** command and by input of the string

```
L(y):=DIF(y,x,3)+DIF(y,x,2)+DIF(y,x)+y
```

The output display is

$$5:\quad L(y) := \left[\frac{d}{dx}\right]^3 y + \left[\frac{d}{dx}\right]^2 y + \frac{d}{dx} y + y$$

Now, this differential operator can be evaluated for y_p through **Author** of the expression

```
AUTHOR expression: l(axê^-x+bx+c)_

Enter expression
User                         Free:100%              Derive Algebra
```

The working area display of this command and the result obtained by executing **Simplify** are shown.

$$6:\quad L\left(a\,x\,\hat{e}^{-x} + b\,x + c\right)$$

$$7:\quad 2\,a\,\hat{e}^{-x} + b\,x + b + c$$

Equating like coefficients results in three simultaneous equations: $2A = 1$, $B = k$, and $B + C = 0$. Their solution is obviously $A = 1/2$, $B = k$, and $C = -k$. If a more complicated system of equations for A, B, and C had to be solved, Derive could have been used to solve the system of equations for these unknowns (see Section 1.8).

The general solution, y_g, is the sum $y_c + y_p$, and can be entered as a defined operator function through **Author** and by input of

```
y(x):=aAlt-e^-x+b cosx+c sinx+0.5xAlt-e^-x=kx-k
```

This operator displays as

```
                   -x                                  -x
10:  Y (x) := a ê     + b COS (x) + c SIN (x) + 0.5 x ê     + k x - k
```

Because of Derive's restriction on subscripts, a, b, and c have been used in place of c_1, c_2, and c_3. In order to check y_g in the original equation, simply **Author** and **Simplify** the expression `L(y(x))`. The output is

```
11:  L (Y (x))

      -x
12:  ê    + k x
```

which shows that the operations on y_g contained in the left side of the equation yield an expression that is equal to the forcing term on the right side, therefore y_g satisfies the equation.

Even before the constants a, b, and c are determined, the value of k that maximizes the solution for $x > 1$ can be determined. The parameter k appears in the solution in the form $kx - k$. For $x > 1$ and $-8 \leq k \leq 8$, $k = 8$ maximizes this expression and, therefore, the solution. Now, $y(x)$ must be redefined with 8 in place of k by executing **Author** and entering

```
AUTHOR expression: Y(x):=aê^(-x)+bCOS(x) + c SIN(x) + 0.5x ê^(-x) + 8 x - 8_

Enter expression
User                                   Free:100%             Derive Algebra
```

Implementing the initial conditions to find values for a, b, and c can also be done easily now that $y(x)$ is a defined function. Simply, **Author** the command `y(1)=1` and select **Simplify** to obtain the first condition. The output is

```
15:  Y (1) = 1

           1    -1
16:  [a + ---] ê    + b COS (1) + c SIN (1) = 1
           2
```

The second condition ($y'(1) = 1/4$) is input by the in-line command `DIF(y(x),x) = 1/4`. **Simplify** results in the following display.

```
      d             1
17:   -- Y (x) =  ---
      dx            4

                         -x
        (x + 2 a - 1) ê                                       1
18:  - ------------------- + c COS (x) - b SIN (x) + 8 =  ---
               2                                              4
```

Manage, Substitute the value 1 for x to get

```
                         -1
        (1 + 2 a - 1) ê                                       1
19:  - ------------------- + c COS (1) - b SIN (1) + 8 =  ---
               2                                              4
```

The third and last condition ($y''(1) = -1/2$) is implemented in a similar manner. The in-line command is `DIF(y(x),x,2) = -1/2`. Once again 1 is substituted for x using the **Manage, Substitute** command. The resulting equations are

```
       ┌d ┐2            1
20:    │──│  Y (x) = - ───
       └dx┘             2

                         -x
       (x + 2 a - 2) ê                                     1
21:    ──────────────────── - b COS (x) - c SIN (x) = - ───
                 2                                         2

                         -1
       (1 + 2 a - 2) ê                                     1
22:    ──────────────────── - b COS (1) - c SIN (1) = - ───
                 2                                         2
```

The three equations for the initial conditions can be solved by placing them as components into a vector. One way to do this is to **Author** the expression **[#m,#n,#p]**, where m, n, and p are the Derive statement numbers of the equations. For the screen displays shown above, the expression would be **[#22,#32,#42]**. Since we will be satisfied with decimal approximations for the coefficients in the final solution, **approX** this expression before executing the **soLve** command. The output of these steps is

```
                                                                    -1
       ┌┌    1 ┐ -1                                 (1 + 2 a - 1) ê
23:    ││a + ──│ ê   + b COS (1) + c SIN (1) = 1, - ──────────────────── + c COS (
       └└    2 ┘                                            2

       ┌          -13              12             12             12
24:    └2.82181 10    (1.30369 10   a + 1.91473 10   b + 2.98202 10   c + 6.51849

       ┌    1003533115        405361674185809978584145672907З           2842417686
25:    │a = ──────────, b = ────────────────────────────────, c = - ──────────
       └    1476716814        612617947628394772877242690500            7948717870

26:    [a = 0.679570, b = 6.61687, c = -3.57594]
```

Now these values for the constants can be substituted into y_g to produce the solution

$$y = 0.679575e^{-x} + 6.61688\cos x - 3.57594\sin x + 0.5xe^{-x} + 8x - 8.$$

To plot this solution, **Author** it into the work area. The various parts can be pulled into the work area using the highlight and the **F3** key. Before issuing the **Plot** command, insure the **Mode** in the **Options, Display** is set to **Graphics**. Execute the menu command **Plot**. This sets up a 2-D plotting screen. Change the scale so the tick marks in the x- and y-directions are more than the default of 1 unit. Since there is an exponential function of x in the solution, it may be wise to have a larger y-scale. Execute the **Scale** command and try an x-scale of 2 and a y-scale of 25. Then issue **Plot** to get the following screen:

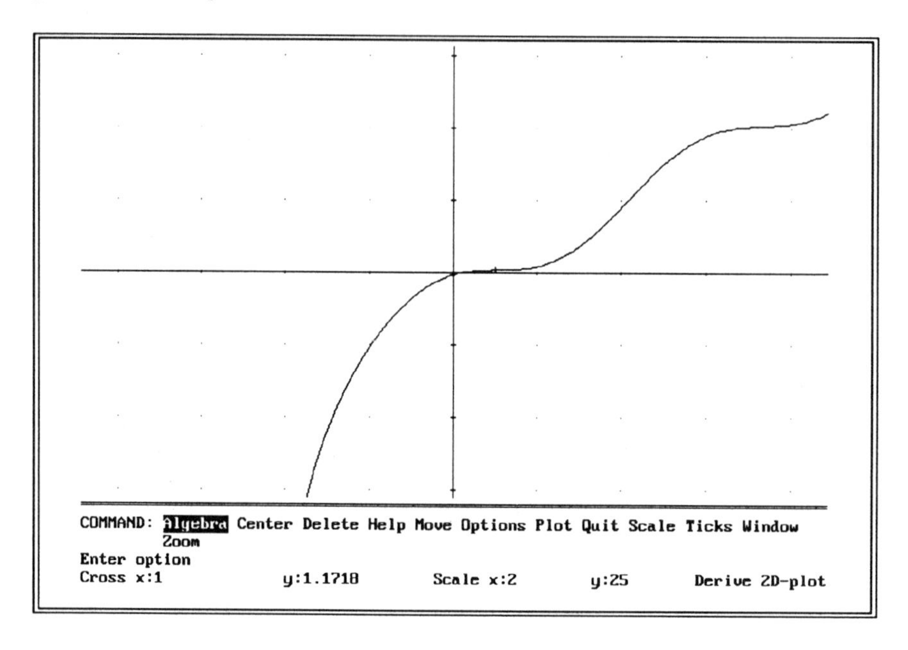

Using the **Zoom** and movable cross with the **Center** command enables the plot range to be adjusted to any region of interest. Don't be impatient, it takes a minute or two for the plot of this function to be completed.

Example 2.6: Linear Algebra

Mathematics is the science which draws necessary conclusions.

—Benjamin Peirce in *Linear Associative Algebra* [1870]

Subject: Solving Systems of Linear Algebraic Equations

Problem: When solving differential equations, there are many times a solution to a linear system of algebraic equations is needed. This is an obvious requirement when working with systems of differential equations. However, this situation is also encountered in partial fraction decomposition, in implementing initial and boundary conditions for higher order equations, and on numerous other occasions. There are 2 ways to solve systems of linear equations using Derive. The following examples demonstrate both ways along with a few subtle extras.

Solution: For the first example, let's find the solution to the 4x4 linear system

$$2x + y + 2z - 3w = 0$$
$$4x + y + z + w = 15$$
$$6x - y - z - w = 5$$
$$4x - 2y + 3z - w = 2$$

One way to solve this is to **Author** the four equations as components of a vector in Derive's work area. The expression is entered and displayed (truncated) as

```
AUTHOR expression: [2x+y+2z-3w=0,4x+y+z+w=15,6x-y-z-w=5,4x-2y+3z-w=2]_

Enter expression
                                   Free:100%            Derive Algebra
```

```
1:   [2 x + y + 2 z - 3 w = 0, 4 x + y + z + w = 15, 6 x - y - z - w = 5, 4 x -
```

Then, **soLve** this expression to obtain the solution

```
2:    [x = 2, y = 3, z = 1, w = 3]
```

The answer is produced, very conveniently, in less than one second.

Let's solve the same system using matrices with Derive. If the above system is written as a matrix-vector system $\mathbf{A}\,\vec{x} = \vec{f}$, then the solution vector $\vec{x}$ can be determined by $\mathbf{A}^{-1}\vec{f}$. For this problem the solution is found by the evaluation of

$$\begin{bmatrix} 2 & 1 & 2 & -3 \\ 4 & 1 & 1 & 1 \\ 6 & -1 & -1 & -1 \\ 4 & -2 & 3 & -1 \end{bmatrix}^{-1} \begin{bmatrix} 0 \\ 15 \\ 5 \\ 2 \end{bmatrix}.$$

This expression is entered into the working area through the **Author** command by typing

```
[[2,1,2,-3],[4,1,1,1],[6,-1,-1,-1], [4,-2,3,-1]]^(-1).[0,15,5,2]
```

Don't forget the dot (.) between the matrix and the vector. Derive uses the dot as the matrix multiplication symbol. Derive displays this expression as

```
     ┌ 2   1   2  -3 ┐-1
     │               │
     │ 4   1   1   1 │
3:   │               │   · [0, 15, 5, 2]
     │ 6  -1  -1  -1 │
     │               │
     └ 4  -2   3  -1 ┘
```

and the **Simplify** command gives the answer

```
4:   [2, 3, 1, 3]
```

These nice integer answers are typical of textbook problems, but such clean results are not expected very often in realistic problems. Not only that, but also solutions to linear systems with integer coefficients can result in quite messy real numbers. For instance, let's change the last equation in the above system to $2x - 2y + 3z - w = 2/7$. To do this easily, highlight the vector of equations previously entered and use the **Author** command and **F3** key to bring that expression into the working area. The **Ctrl-s** and **Ctrl-d** keys provide movement of the cursor in the expression so the coefficients in the last expression can be changed. This is an instance when smart keystroking saves alot of time. No one wants to retype all 4 of these equations. **Enter** this expression and **soLve** to obtain the solution

```
6:  [2 x + y + 2 z - 3 w = 0, 4 x + y + z + w = 15, 6 x - y - z - w = 5, 2 x -

                  361        211        457
7:  [x = 2, y = ───── , z = ───── , w = ─────]
                  147        147        147
```

in only 1.2 seconds or so depending on the speed of the computer being used. These rational numbers can be converted to approximate 6-digit decimal equivalents with the **approX** command to get

```
                  361        211        457
7:  [x = 2, y = ───── , z = ───── , w = ─────]
                  147        147        147

8:  [x = 2, y = 2.45578, z = 1.43537, w = 3.10884]
```

Derive has no problem handling parameters in the equations. For instance, let's change the 2/7 in the last equation to the parameter α (it's time to see Derive do Greek letters). Follow the keyboarding steps above to obtain the previous set of equations and change the 2/7 to α. To get Derive to use the Greek letter α, type **ALT-a**. **soLve** this expression with the solve variables set to x, y, z, and w to obtain

```
10:  [2 x + y + 2 z - 3 w = 0, 4 x + y + z + w = 15, 6 x - y - z - w = 5, 2 x -

                     5 α - 53        4 α + 29        α + 65
11:  [x = 2, y = - ────────── , z = ────────── , w = ────────]
                       21              21              21
```

The second example shows when and how matrix manipulation can be used effectively to solve systems of equations. Sometimes the problem requires solutions for several systems with different right-hand sides. The following matrix equation can represent 4 different 3x3 systems of equations. In matrix form, this example problem is

$$\begin{bmatrix} 1 & -1 & 1 \\ 1 & -2 & -2 \\ 2 & 1 & 3 \end{bmatrix} \begin{bmatrix} x & u & m & r \\ y & v & n & s \\ z & w & p & t \end{bmatrix} = \begin{bmatrix} 3 & 1 & -6 & 3/2 \\ 0 & -1 & 1 & 0 \\ 4 & 3 & -9 & 11/2 \end{bmatrix}.$$

A way to enter matrices into Derive is through the **Declare, Matrix** command, which prompts the user for matrix size (rows and columns) and the elements of the matrix. Let's enter the two matrices given above containing numbers as separate working expressions in Derive using the **Declare, Matrix** command. The first matrix has 3 rows and 3 columns. The second matrix has 3 rows and 4 columns. The result of the input of the elements in these matrices is the following screen display of the two matrices.

```
          ┌ 1  -1   1 ┐
12:       │ 1  -2  -2 │
          └ 2   1   3 ┘

          ┌                3 ┐
          │ 3   1  -6    ─── │
          │                2 │
13:       │ 0  -1   1     0  │
          │               11 │
          │ 4   3  -9    ─── │
          └                2 ┘
```

The solution matrix is obtained by multiplying the inverse of the 3x3 matrix times the 3x4 matrix. This can be accomplished with by **Author** of the command **#n^(-1).#k**, where **n** and **k** are the numbers of the expression for these matrices in the work area. Don't forget the dot between the matrices. For the above Derive screen, the command is **#12^(- 1).#13**. This displays as

$$14: \begin{bmatrix} 1 & -1 & 1 \\ 1 & -2 & -2 \\ 2 & 1 & 3 \end{bmatrix}^{-1} \cdot \begin{bmatrix} 3 & 1 & -6 & \frac{3}{2} \\ 0 & -1 & 1 & 0 \\ 4 & 3 & -9 & \frac{11}{2} \end{bmatrix}$$

`Simplify` this expression to obtain the solution matrix

$$15: \begin{bmatrix} \frac{1}{2} & \frac{1}{2} & -1 & 2 \\ -\frac{9}{8} & \frac{1}{8} & 2 & \frac{3}{4} \\ \frac{11}{8} & \frac{5}{8} & -3 & \frac{1}{4} \end{bmatrix}$$

Example 2.7: Characteristic Values

Nothing makes a little knowledge as dangerous as examination time.

—Anonymous Saying

Subject: Finding Characteristic Values and Characteristic Vectors (Eigenvalues and Eigenvectors)

Problem: One of the common tasks in working with systems of linear equations is to find the characteristic values and characteristic vectors of a matrix. Derive can help tremendously in these tasks. In fact, there are several easy ways to use Derive, which are demonstrated in the following examples.

Solution: The first task is to find the characteristic values and vectors of the rather messy 3x3 matrix **A**, where

$$\mathbf{A} = \begin{bmatrix} 3 & 2/9 & -1 \\ 2 & 2 & 3 \\ 1 & 1/6 & 9 \end{bmatrix}.$$

First, enter the matrix **A** into the work area either directly or using the `Declare, Matrix` command (see Section 2.6). Then, `Author` the expression `EIGENVALUES #n`, where `n` is the expression number for the matrix. The Derive display as a result of these steps for this problem is shown on the next page.

$$1: \begin{bmatrix} 3 & \frac{2}{9} & -1 \\ 2 & 2 & 3 \\ 1 & \frac{1}{6} & 9 \end{bmatrix}$$

$$2: \text{EIGENVALUES} \begin{bmatrix} 3 & \frac{2}{9} & -1 \\ 2 & 2 & 3 \\ 1 & \frac{1}{6} & 9 \end{bmatrix}$$

Simplify to yield the solutions for the three eigenvalues using the default variable w.

$$3: \left[w = \frac{5}{3},\ w = \frac{37}{6} - \frac{\sqrt{271}}{6},\ w = \frac{\sqrt{271}}{6} + \frac{37}{6} \right]$$

To find the characteristic vectors (eigenvectors) for each of these eigenvalues, the matrix equation $(\mathbf{A} - w\mathbf{I})\vec{k} = 0$, where $\mathbf{I}$ is the identity matrix, is solved for $\vec{k}$. The expression $(\mathbf{A} - w\mathbf{I})$ is entered by highlighting matrix **A** and using **F3** to bring it into the work area and then appending `- w IDENTITY_MATRIX(3)`. The input line and its displays are shown.

```
AUTHOR expression: [[3, 2/9, -1], [2, 2, 3], [1, 1/6, 9]] -widentity_matrix(3)_

Enter expression
User                                  Free:100%            Derive Algebra
```

$$4: \quad \begin{bmatrix} 3 & \frac{2}{9} & -1 \\ 2 & 2 & 3 \\ 1 & \frac{1}{6} & 9 \end{bmatrix} - w \ \mathrm{IDENTITY\_MATRIX}(3)$$

The 3 in the `IDENTITY_MATRIX(3)` command produces a 3x3 identity matrix. **`Simplify`** this expression to get the matrix (**A** - w**I**).

$$5: \quad \begin{bmatrix} 3 - w & \frac{2}{9} & -1 \\ 2 & 2 - w & 3 \\ 1 & \frac{1}{6} & 9 - w \end{bmatrix}$$

Complete the left-hand side of the expression by dotting this matrix with $\vec{k}$. Let's use

$$\vec{k} = \begin{bmatrix} a \\ b \\ c \end{bmatrix}$$

for the Derive input. Once again, highlight the previous matrix expression and use the **`Author`** command and **F3** to move it into the work area. Then append the rest of the expression `.[a,b,c]` onto the expression in the active work area. Derive shows the matrix-vector multiplication by

```
         [          2                ]
         [ 3 - w   ———     -1        ]
         [          9                ]
         [                           ]
    6:   [   2     2 - w    3        ] · [a, b, c]
         [                           ]
         [          1                ]
         [   1     ———    9 - w      ]
         [          6                ]
```

Simplify to obtain the vector of expressions

```
                     2 b                                              b
 7:   [a (3 - w) + ——— - c, 2 a + b (2 - w) + 3 c, a + ——— + c (9 - w)]
                      9                                               6
```

Before solving for the eigenvector components a, b, and c, the value of one of the eigenvalues needs to be substituted for w. This is done with the **Manage, Substitute** command. Let's substitute 5/3 in for w first. Derive prompts the user for values for each of the variables in the expression. Just hit **Return** or **Enter** for the variables a, b, and c to signify they remain as unknown variables and 5/3 **Enter** when prompted for w.

By issuing the **soLve** command, Derive assumes each expression is set to 0 and solves for a, b, and c. The results of the operation are the following vector components for the eigenvector.

```
 9:   [a = @1, b = - 6 @1, c = 0]
```

Notice the value @1 in two of the vector components. Derive uses the values @n, n integer, to stand for arbitrary constants. So any value can be substituted for @1 to produce an eigenvector associated with the eigenvalue 5/3.

The other two eigenvectors for the eigenvalues $37/6-\sqrt{271}/6$ and $37/6+\sqrt{271}/6$ are computed following the same steps starting with the **Manage, Substitute** command and substituting these values for w. The highlight and **F3** key can be used to get these values when the prompt for w is given. Then the eigenvector are found using the **soLve** command. The Derive expressions for $w = 37/6 - \sqrt{271}/6$ are as follows:

$$10: \left[a\left[3-\left[\frac{37}{6}-\frac{\sqrt{271}}{6}\right]\right]+\frac{2\ b}{9}-c,\ 2\ a+b\left[2-\left[\frac{37}{6}-\frac{\sqrt{271}}{6}\right]\right]+3\ c,\right.$$

$$11: \left[a=@2,\ b=@2\left[\frac{9\sqrt{271}}{85}-\frac{66}{85}\right],\ c=@2\left[\frac{97\sqrt{271}}{510}-\frac{1703}{510}\right]\right]$$

Similarly, the displays for $w = 37/6 + \sqrt{271}/6$ are as follows:

$$12: \left[a\left[3-\left[\frac{\sqrt{271}}{6}+\frac{37}{6}\right]\right]+\frac{2\ b}{9}-c,\ 2\ a+b\left[2-\left[\frac{\sqrt{271}}{6}+\frac{37}{6}\right]\right]+3\ c,\right.$$

$$13: \left[a=@3,\ b=-@3\left[\frac{9\sqrt{271}}{85}+\frac{66}{85}\right],\ c=-@3\left[\frac{97\sqrt{271}}{510}+\frac{1703}{510}\right]\right]$$

The next task is to find the characteristic values and vectors of a 4x4 matrix **B**, where

$$\mathbf{B} = \begin{bmatrix} 1 & 0 & 0 & 0 \\ 2 & -1 & 4 & 2 \\ 3 & 4 & -1 & -2 \\ 4 & 0 & 0 & 6 \end{bmatrix}.$$

Don't be misled, Derive does not like to work with 4x4 matrices because the characteristic polynomial is fourth degree. Some fourth degree polynomials don't lend themselves to easy solutions. This particular matrix is nice in this regard. Derive doesn't mind complex numbers for eigenvalues. See Sections 3.9 for a problem with complex eigenvalues. The moral of the story is that even systems as small as 4x4 can be very difficult.

After entering the matrix, the eigenvalues can be solved for by use of the `CHARPOLY` function and `soLve` menu command or the `EIGENVALUES` function. For some 4x4 matrices Derive may work a long time (2-4 minutes) to get a solution or even give up completely. Derive indicates it has given up by returning the same expression it started with. The `CHARPOLY`, `Simplify` and `soLve` commands and their results for this problem are shown.

```
                  ┌ 1   0   0   0 ┐
                  │               │
                  │ 2  -1   4   2 │
15:   CHARPOLY    │               │
                  │ 3   4  -1  -2 │
                  │               │
                  └ 4   0   0   6 ┘

                                2
16:   (w - 6) (w - 1) (w  + 2 w - 15)

17:   w = 1

18:   w = 3

19:   w = -5

20:   w = 6
```

The eigenvectors are found by solving for $\vec{k}$ in the equation $(\mathbf{B} - w\mathbf{I})\vec{k} = 0$, where $\mathbf{I}$ is the identity matrix and w is an eigenvalue. Follow the steps of the previous example with $w = 6$. This time the **IDENTITY_MATRIX(4)** function needs to be used, and the vector $\vec{k}$ needs four components a, b, c, and d. The display of this sequence of operations is as follows:

```
      ┌ 1   0   0   0 ┐
      │               │
      │ 2  -1   4   2 │
21:   │               │ - w IDENTITY_MATRIX (4)
      │ 3   4  -1  -2 │
      │               │
      └ 4   0   0   6 ┘

      ┌ 1 - w     0        0       0   ┐
      │                                │
      │   2     -w - 1     4       2   │
22:   │                                │
      │   3       4      -w - 1   -2   │
      │                                │
      └   4       0        0     6 - w ┘

      ┌ 1 - w     0        0       0   ┐
      │                                │
      │   2     -w - 1     4       2   │
23:   │                                │ · [a, b, c, d]
      │   3       4      -w - 1   -2   │
      │                                │
      └   4       0        0     6 - w ┘
```

```
24:  [a (1 - w), 2 a - b (w + 1) + 4 c + 2 d, 3 a + 4 b - c (w + 1) - 2 d, 4 a +

25:  [a (1 - 6), 2 a - b (6 + 1) + 4 c + 2 d, 3 a + 4 b - c (6 + 1) - 2 d, 4 a +

                                          11 @4
26:  [a = 0, b = @4, c = -@4, d = ------]
                                            2
```

The other three eigenvectors are produced with the same procedure. The displays for these computations for $= 1, 3, -5$, respectively, are as follows:

```
27:  [a (1 - 1), 2 a - b (1 + 1) + 4 c + 2 d, 3 a + 4 b - c (1 + 1) - 2 d, 4 a +

                      8 @5           9 @5          4 @5
28:  [a = @5, b = - ------, c = - ------, d = - ------]
                       5             10             5
```

```
29:  [a (1 - 3), 2 a - b (3 + 1) + 4 c + 2 d, 3 a + 4 b - c (3 + 1) - 2 d, 4 a +

30:  [a = 0, b = @6, c = @6, d = 0]
```

```
31:  [a (1 - -5), 2 a - b (-5 + 1) + 4 c + 2 d, 3 a + 4 b - c (-5 + 1) - 2 d, 4

32:  [a = 0, b = @7, c = -@7, d = 0]
```

The final problem in this section demonstrates one of things that can happen when there are repeated eigenvalues. Let's work with the 3x3 matrix **C**, where

$$\mathbf{C} = \begin{bmatrix} 1 & 1 & 1 \\ 2 & 1 & -1 \\ -3 & 2 & 4 \end{bmatrix}.$$

Finding the eigenvalues produces the surprise that there is only one. By default, Derive labels it w as shown in this computation.

```
                       [  1  1   1 ]
34:  EIGENVALUES       [  2  1  -1 ]
                       [ -3  2   4 ]

35:  [w = 2]
```

Through the following sequence of steps, Derive builds $(\mathbf{C} - w\mathbf{I})\vec{k}$ with the value 2 substituted for w.

```
      [  1  1   1 ]
36:   [  2  1  -1 ] - w IDENTITY_MATRIX (3)
      [ -3  2   4 ]

      [ 1 - w    1      1   ]
37:   [   2    1 - w   -1   ]
      [  -3      2    4 - w ]

      [ 1 - w    1      1   ]
38:   [   2    1 - w   -1   ] · [a, b, c]
      [  -3      2    4 - w ]

39:  [a (1 - w) + b + c, 2 a + b (1 - w) - c, - 3 a + 2 b + c (4 - w)]

40:  [a (1 - 2) + b + c, 2 a + b (1 - 2) - c, - 3 a + 2 b + c (4 - 2)]
```

Then, **soLve** for the first eigenvector to obtain

```
41:  [a = 0, b = @8, c = -@8]
```

This is the only linearly independent eigenvector for this matrix. The next step is to compute the first generalized eigenvector by solving for $\vec{m}$ in $(\mathbf{C} - w\mathbf{I})\vec{m} = \vec{k}$. In Derive let

$$\vec{m} = \begin{bmatrix} d \\ e \\ f \end{bmatrix}.$$

Author the expression $(\mathbf{C} - w\mathbf{I})\vec{m} - \vec{k}$ using the highlight and **F3** key where appropriate. A specific value should be substituted for the arbitrary constant @8. Often the value 1 is the simplest to use, so let's use it in the **Manage, Substitute** command. Then **soLve** to obtain the vector components.

```
42:  [a = 0, b = 1, c = -1]

     ⎡⎡ 1  1   1 ⎤                     ⎤
43:  ⎢⎢ 2  1  -1 ⎥ - w IDENTITY_MATRIX (3)⎥ · [d, e, f] - [0, 1, -1]
     ⎣⎣-3  2   4 ⎦                     ⎦

44:  [d (1 - w) + e + f, 2 d + e (1 - w) - f - 1, - 3 d + 2 e + f (4 - w) + 1]

45:  [d (1 - 2) + e + f, 2 d + e (1 - 2) - f - 1, - 3 d + 2 e + f (4 - 2) + 1]

46:  [d = 1, e = @9, f = 1 - @9]
```

Next, the second generalized eigenvector is computed as $\vec{p}$ in the matrix-vector equation $(\mathbf{C} - w\mathbf{I})\vec{p} = \vec{m}$. Let

$$\vec{p} = \begin{bmatrix} g \\ h \\ i \end{bmatrix}.$$

Just as the previous generalized eigenvector was found, **Author** the expression $(\mathbf{C} - w\mathbf{I})\vec{p} - \vec{m}$. Substitute 1 for the arbitrary constant @9. Then **soLve** to obtain the vector components.

```
     ⎡⎡ 1  1   1 ⎤                     ⎤
47:  ⎢⎢ 2  1  -1 ⎥ - w IDENTITY_MATRIX (3)⎥ · [g, h, i] - [1, 1, 0]
     ⎣⎣-3  2   4 ⎦                     ⎦

48:  [g (1 - w) + h + i - 1, 2 g + h (1 - w) - i - 1, - 3 g + 2 h + i (4 - w)]

49:  [g (1 - 2) + h + i - 1, 2 g + h (1 - 2) - i - 1, - 3 g + 2 h + i (4 - 2)]

50:  [g = 2, h = @1, i = 3 - @1]
```

When a system of differential equations has repeated eigenvalues that produce generalized eigenvectors, the solution takes on special form. See Sections 2.8 and 3.8 for more on the use of eigenvalues and eigenvectors in solving systems of differential equations.

Example 2.8: System of Differential Equations

CAS is not just an aid to problem solving, students can be guided in the use of CAS to discover fundamental mathematical concepts for themselves.

— Nancy Baxter and Priscilla Laws [1989]

Subject: Solving a Nonhomogeneous System of Differential Equations using Variation of Parameters

Problem: Find a general solution of

$$x_1' = x_1 - x_2 + \frac{e^{-t}}{1+t^2},$$

$$x_2' = 2x_1 - 2x_2 + \frac{2e^{-t}}{1+t^2},$$

with $x_1(1) = 0$ and $x_2(1) = 1$.

Solution: The matrix-vector form of the equation is

$$\vec{X}' = \begin{bmatrix} 1 & -1 \\ 2 & -2 \end{bmatrix} \vec{X} + \vec{F}(t),$$

$$\text{where } \vec{X} = \begin{bmatrix} x_1 \\ x_2 \end{bmatrix} \text{ and } \vec{F}(t) = \begin{bmatrix} \frac{e^{-t}}{1+t^2} \\ \frac{2e^{-t}}{1+t^2} \end{bmatrix}.$$

The eigenvalues of the matrix are found using Derive by **Author** of the in-line command

```
EIGENVALUES [[1,-1],[2,-2]]
```

which displays as

```
1:   EIGENVALUES [ 1  -1 ]
                 [ 2  -2 ]
```

Select the command **Simplify** to obtain the Derive display

```
2:   [w = 0, w = -1]
```

which means that the two eigenvalues are $\lambda_1 = -1$ and $\lambda_2 = 0$. Find the eigenvectors in the form $\begin{bmatrix} j \\ k \end{bmatrix}$ by solving

$$\begin{bmatrix} 1-\lambda & -1 \\ 2 & -2-\lambda \end{bmatrix} \begin{bmatrix} j \\ k \end{bmatrix} = 0$$

For this problem, the solutions for j and k are simple. The calculations using Derive are done for $\lambda = -1$ and 0, respectively, (Derive uses w instead of λ) by issuing **Author** and **soLve**. The resulting display is as follows:

```
3:   [ 1  -1 ]  - w IDENTITY_MATRIX (2)
     [ 2  -2 ]

4:   [ 1 - w     -1    ]
     [   2     -w - 2  ]

5:   [ 1 - w     -1    ]  · [a, b]
     [   2     -w - 2  ]

6:   [a (1 - w) - b, 2 a - b (w + 2)]

7:   [a (1 - 0) - b, 2 a - b (0 + 2)]

8:   [a = @2, b = @2]

9:   [a (1 - -1) - b, 2 a - b (-1 + 2)]

10:  [a = @3, b = 2 @3]
```

Substituting 1 for j in both eigenvectors gives

$$\vec{K_1} = \begin{bmatrix} 1 \\ 2 \end{bmatrix} \text{ and } \vec{K_2} = \begin{bmatrix} 1 \\ 1 \end{bmatrix}.$$

From this result, the complementary solution is written as

$$\vec{X_c} = a\begin{bmatrix} 1 \\ 1 \end{bmatrix} + b\begin{bmatrix} 1 \\ 2 \end{bmatrix} e^{-t}, \text{with } a \text{ and } b \text{ arbitrary constants,}$$

and the fundamental matrix, Φ, is

$$\begin{bmatrix} 1 & e^{-t} \\ 1 & 2e^{-t} \end{bmatrix}.$$

This matrix is used to solve for the particular solution by using the variation of parameters formula

$$\vec{X_p} = \Phi(t) \int \Phi(t)^{-1} \vec{F}(t) dt.$$

First, use Derive to find Φ^{-1}, by use of the **Author** command and entering `[[1,Alt-e^-t], [1,2Alt-e^-t]]^-1`. Notice that e is entered by pressing the **Alt** key and the letter **e** key, simultaneously. Derive displays it as $\hat{e}$. **Simplify** this expression to find Φ^{-1}. At this point, the Derive display and authored expression are as follows:

```
           -t   -1
    [ 1   ê    ]
11: [          ]
    [        -t]
    [ 1  2 ê   ]

    [   2    -1 ]
12: [           ]
    [    t    t ]
    [ - ê    ê  ]

AUTHOR expression: [[2, -1], [- ê^t, ê^t]].[ê^-t/(1+t^2),2ê^-t/(1+t^2)]_

Enter expression
Solve(9)                        Free:89%                 Derive Algebra
```

Multiply Φ^{-1} by $\vec{F}(t)$. This is best done by building from the previous expression for Φ^{-1} using the **Author** command and **F3**, then dotting this with $\vec{F}(t)$. **Simplify** this expression and integrate the resulting vector using the **Calculus, Integrate** menu command. The integration must be done with respect to t with no explicit limit of integration specified. Just press **Enter** when queried for the limits. The resulting display upon the execution of the **Simplify** command is shown.

```
        ┌  2   -1 ┐   ┌   -t          -t  ┐
        │         │   │  ê         2 ê    │
13:     │   t   t │ · │ ───────,  ─────── │
        └ - ê   ê ┘   │      2          2 │
                      └ 1 + t      1 + t  ┘

        ┌      1    ┐
        │0, ─────── │
14:     │      2    │
        └     t  + 1┘

        ┌      1    ┐
      ⌠ │0, ─────── │ dt
15:   ⌡ │      2    │
        └     t  + 1┘

16:   [0, ATAN (t)]
```

Following the equation for the variation of parameters given above, we multiply this expression by Φ to obtain $\vec{X}_p$, which is done by *Authoring* the Derive expression

```
AUTHOR expression: [[1, ê^(-t)], [1, 2 ê^(-t)]] · [0, ATAN (t)]_

Enter expression
User                                  Free:89%                Derive Algebra
```

Perform the operation through the command **Simplify**. The display for this is shown.

```
        ┌      -t ┐
        │ 1   ê   │
17:     │         │ · [0, ATAN (t)]
        │       -t│
        └ 1  2 ê  ┘

        ┌ -t              -t          ┐
18:     └ê   ATAN (t), 2 ê   ATAN (t) ┘
```

The general solution, $\vec{X}_g$, is $\vec{X}_c + \vec{X}_p$ and can now be written as

$$\vec{X}_g = a\begin{bmatrix} 1 \\ 1 \end{bmatrix} + b\begin{bmatrix} 1 \\ 2 \end{bmatrix} e^{-t} + \begin{bmatrix} 1 \\ 2 \end{bmatrix} e^{-t}\tan^{-1} t.$$

The arbitrary constants a and b are determined by substitution of the initial conditions into $\vec{X}_g$ and solving for a and b. To do this with Derive, the two equations are entered as a vector of equations with x used for x_1 and y used for x_2. Then using the menu commands **Manage, Substitute**, the initial values of $t = 1, x = 0$, and $y = 1$ are entered. The display showing these operations is given.

```
                   -t     -t                                 -t       -t
20:  [x = a + b ê   + ê   ATAN (t), y = a + b 2 ê   + 2 ê   ATAN (t)]

                   -1     -1                                 -1       -1
21:  [0 = a + b ê   + ê   ATAN (1), 1 = a + b 2 ê   + 2 ê   ATAN (1)]
```

The values for a and b are obtained by selecting the command **soLve**. Derive displays the result as follows.

```
                               π
22:  [a = -1, b = ê - ───]
                               4
```

Finally, the solution to the initial-value problem is written as

$$x_1 = -1 + (e - \pi/4)e^{-t} + e^{-t}\tan^{-1} t \qquad \text{and}$$

$$x_2 = -1 + 2(e - \pi/4)e^{-t} + 2e^{-t}\tan^{-1} t.$$

To get a geometric perspective of this solution, x_1 and x_2 are plotted using Derive's 2-D plotting capability. Just **Author** the two solutions as equations into the Derive display as shown.

```
                    ┌     π ┐  -t     -t
24:  x = -1 + │ê - ───│ ê    + ê   ATAN (t)
                    └     4 ┘

                    ┌     π ┐    -t       -t
26:  y = -1 + │ê - ───│ 2 ê    + 2 ê   ATAN (t)
                    └     4 ┘
```

Before plotting the two functions, remember to set the plotting parameters to those of your hardware using the **Options, Display** menu commands. A plot showing the local behavior near the origin is obtained with the scale of the tick marks set to the default value of 1.0 in both the x (horizontal) and y (vertical) directions. This plot is as follows:

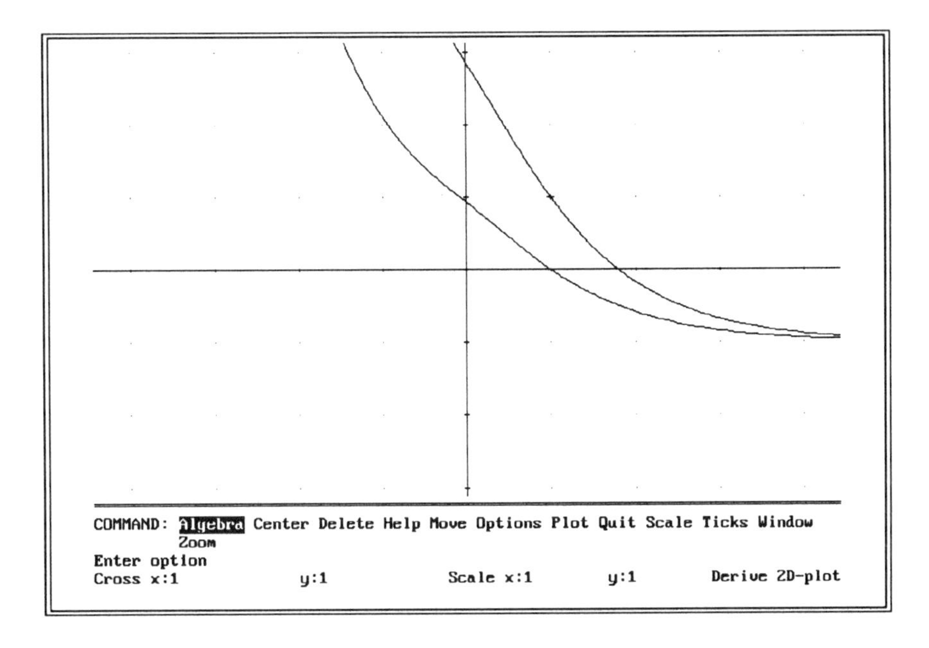

A plot of the global behavior is obtained by changing the scale to 10 for every tick mark on both axes. Change the scale by using the `Scale` command and entering 10 for both the x- and y–scale. Use the **Tab** key to move between these selections in the menu. This plot of the global behavior of the solution components is shown below.

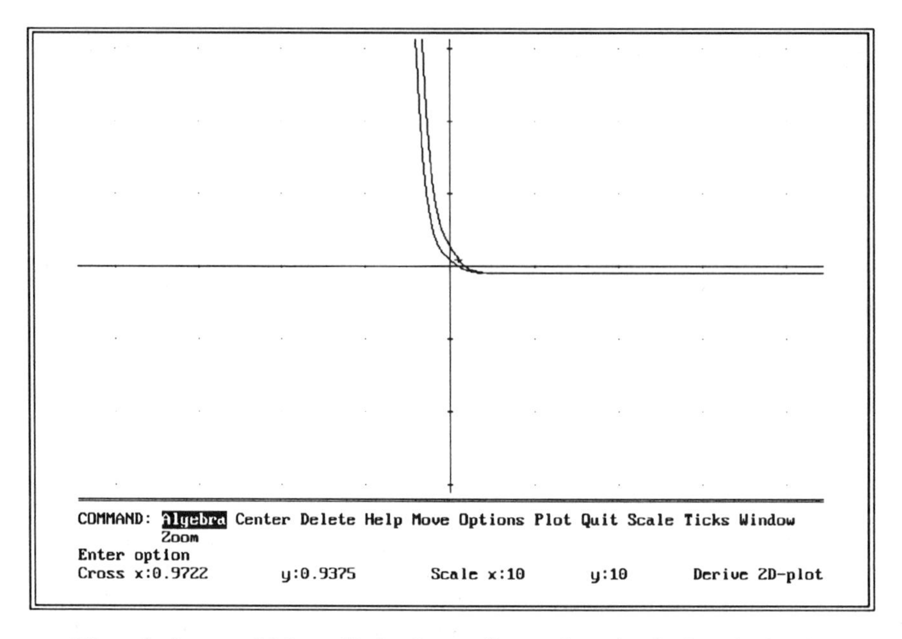

The window could be split horizontally to show both the algebraic and graphical forms of the solutions. To do this, go back to the algebra screen by selecting **Algebra**. Execute the **Window, Split** commands and select a **Horizontal** split at line #13. Issue the **Plot** command. In order to make a better plot for this window, change the x−scale to 1 and the y−scale to 25. The following screen image is the final result of these operations.

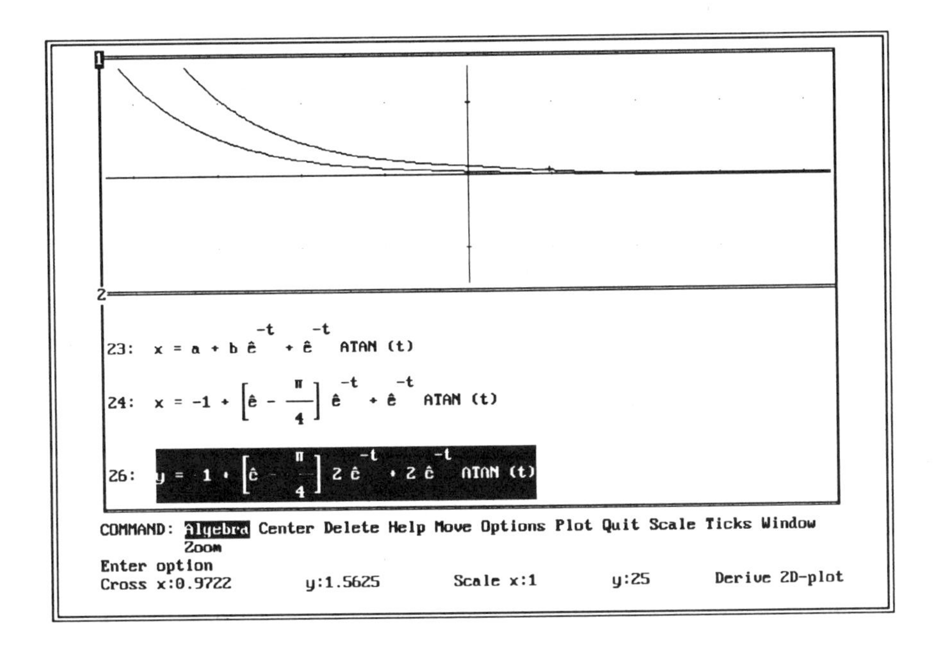

1
2
23: x = a + b ê^-t + ê^-t ATAN (t)
24: x = -1 + [ê - π/4] ê^-t + ê^-t ATAN (t)
26: y = 1 + [ê - π/4] 2 ê^-t + 2 ê^-t ATAN (t)
COMMAND: Algebra Center Delete Help Move Options Plot Quit Scale Ticks Window
Zoom
Enter option
Cross x:0.9722 y:1.5625 Scale x:1 y:25 Derive 2D-plot

Example 2.9: Electrical Circuit

...the aim of exact science is to reduce the problems of nature to the determination of quantities by operations with numbers.

— James Maxwell [1856]

Subject: The Series Electrical Circuit

Problem: The basic, simple-loop electrical circuit consists of several parts, a time-dependent voltage source ($E(t)$), an inductor (L), a resistor (R), and a capacitor (C), arranged in series. Variations of this circuit containing more than one inductor, resistor, and capacitor are found throughout homes. In these simple circuits, L, R, and C represent the loop's total induction, resistance, and capacitance, respectively. The problem is to insure that the electrical current in any loop does not exceed the safe wire capacity which is protected by the capacity of the fuse or circuit breaker.

Solution: The best way to determine the current is to set up a differential model using Kirchoff's laws. By letting $q(t)$ be the time-dependent charge on the capacitor, the model is

$$L\frac{d^2q}{dt^2} + R\frac{dq}{dt} + \frac{1}{C}q = E(t)$$

where L, R, and C are assumed to be constant for this circuit.

The applied voltage for a typical household alternating current can be modeled as

$$E(t) = 110\cos 2t \text{ Volts.}$$

The Derive plot of this voltage is made with the `Author` command and by entering this expression `110 cos(2t)` into the work area and using the `Plot` command. Good parameters for the `Scale` command are x-scale = 5 and y-scale = 50. `Move` the cross to $x = 20$ and $y = 0$ and execute the `Center` command. Don't forget the **Tab** key in order to move the cursor between different parts of the menu. After these parameters are set, then the `Plot` command is issued to begin drawing the graph of the function. The resulting voltage plot is shown.

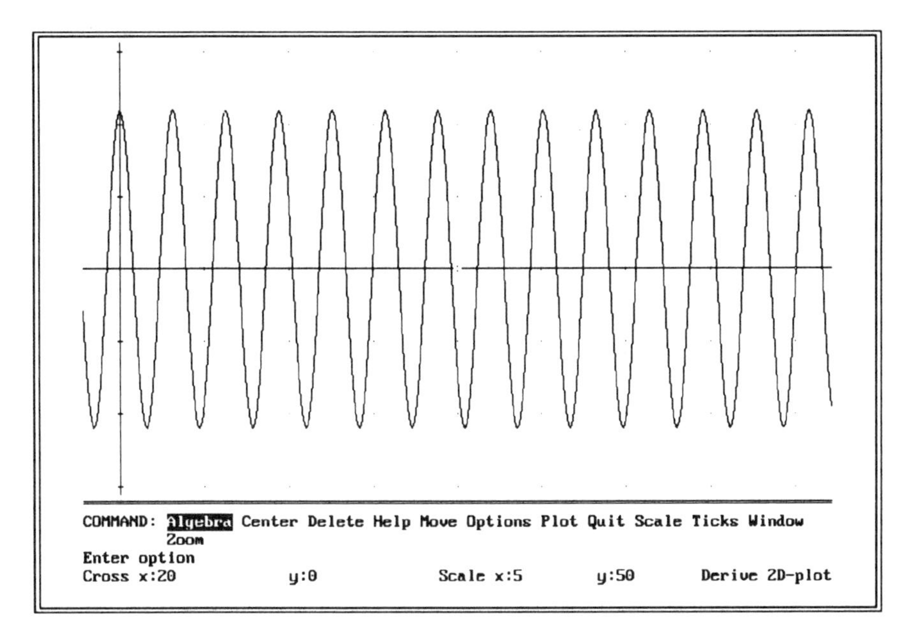

For a circuit with $L = 10$ henrys, $R = 50$ ohms, $C = 1/60$ farads, no initial charge on the capacitor ($q(0) = 0$), no initial current ($q'(0) = 0$), and $E(t)$ applied at $t = 0$ (q measured in volts, t measured in seconds), the equation to be solved is

$$10\frac{d^2q}{dt^2} + 50\frac{dq}{dt} + 60q = 110\cos 2t,$$

with $q(0) = 0$ and $q'(0) = 0$.

To solve this equation with Derive, first use the **Transfer, Merge** command to obtain the utility file ODE2.MTH which has commands available to solve second-order, constant coefficient, nonhomogeneous equations in the form $y'' + ay' + by = r$ (Note: a and b have been substituted for the p and q in file ODE2 because q has already been used as the dependent variable in the equation). Next, place the equation in this standard form

$$\frac{d^2q}{dt^2} + 5\frac{dq}{dt} + 6q = 11\cos 2t$$

to identify $a = 5$ and $b = 6$.

Use the command `LIN2_RED_CCF_DISC(5,6)` and **Simplify** to obtain these following result for the evaluation of the discriminant.

```
36:  LIN2_RED_CCF_DISC (5, 6)

37:  1
```

The positive value of 1 for the discriminant resulting from this command determines the next command in the solution sequence. Therefore, for this case, the command **LIN2_RED_CCF_POS(5,6,t)** computes the complementary solution to the equation. As shown in a table of Section 1.12, if the discriminant had been negative the command to solve the equation would be **LIN2_RED_CCF_NEG** and if the discriminant had evaluated to 0, the command would be **LIN2_RED_CCF_0**.

For this problem, the command and resulting display after **Simplify** are as follows.

```
38:  LIN2_RED_CCF_POS (5, 6, t)

            - 2 t        - 3 t
39:  @1 ê         + @2 ê
```

Using the notation of the commands in utility file ODE2 on this result, the two fundamental solutions are $ux = e^{-2t}$ and $vx = e^{-3t}$. The next command, **LIN2_COMPLETE**, uses the method of variation of parameters to obtain a particular solution to the nonhomogeneous equation. **Author** the command

LIN2_COMPLETE(Alt-e^(-2t), Alt-e^(-3t),11cos(2t),t).

The highlight and **F3** key can help build this command to save keystrokes. Derive's **Simplify** command provides the particular solution as shown in the following display.

```
                        - 2 t    - 3 t
40:  LIN2_COMPLETE (ê       , ê       , 11 COS (2 t), t)

     11 COS (2 t)        55 SIN (2 t)
41:  ─────────────  +  ─────────────
          52                  52
```

The general solution can be assembled from the complementary and particular solutions as

$$q(t) = @1e^{-2t} + @2e^{-3t} + \frac{11\cos 2t}{52} + \frac{55\sin 2t}{52}$$

using Derive's @1 and @2 notation for the arbitrary constants.

The two initial conditions are imposed to find values for @1 and @2 using the command

```
IMPOSE_IC2(t, @1Alt-e^(-2t)+@2Alt-e^(-3t)+11cos(2t)/52+ 55sin(2t)/52,0,0,0)
```
.

Once again, the most efficient way to enter this is by moving the highlight to parts of previous expressions and using the **F3** key to bring these highlighted expressions into the work area. `Simplify` this expression to get the following values for @1 and @2.

```
                   ┌        - 2 t        - 3 t   11 COS (2 t)    55 SIN (2 t)
42:  IMPOSE_IC2    │t, @1 ê      + @2 ê      + ──────────── + ────────────, 0,
                   └                               52              52

     ┌        11        33  ┐
43:  │ @1 = - ──   @2 = ──  │
     └        4         13  ┘
```

The solution to the differential equation can now be written as

$$q(t) = -\frac{11}{4}e^{-2t} + \frac{33}{13}e^{-3t} + \frac{11\cos 2t}{52} + \frac{55\sin 2t}{52}.$$

This is entered into Derive through the use of the `Manage, Substitute` command on the expression for the general solution. Execution of this command results in queries for values for each of the remaining variable in the expression. In this case just press **Enter** for the value for t (indicating no substitution) and the values of -11/4 and 33/13 for values of `@1` and `@2`, respectively. The resulting Derive display for this is as follows.

$$44: \left[-\frac{11}{4}\right] \hat{e}^{-2t} + \frac{33}{13}\hat{e}^{-3t} + \frac{11 \text{COS}(2t)}{52} + \frac{55 \text{SIN}(2t)}{52}$$

Before plotting this solution, it should be checked. Derive can help do this. **Author** the operator for the equation by entering

```
AUTHOR expression: 10 DIF (q, t, 2) + 50 DIF (q, t) + 60 q - 110 COS (2 t)_

Enter expression
User                    D:2.9              Free:89%             Derive Algebra
```

Use the **Manage, Substitute** command to substitute the expression for the solution into the equation for the variable q. To do this, use the highlight and **F3** key to put the equation in the input line of the menu or use the # expression reference number (in this case #44. The display for this resulting command is truncated on the right in the following figure because of its length.

$$46: \ 10 \left[\frac{d}{dt}\right]^2 \left[\left[-\frac{11}{4}\right] \hat{e}^{-2t} + \frac{33}{13}\hat{e}^{-3t} + \frac{11 \text{COS}(2t)}{52} + \frac{55 \text{SIN}(2t)}{52}\right]$$

Simplify to get verification that the function does, in fact, evaluate to 0.

Since the current $i(t)$ is equal to $q'(t)$, use Derive to get an expression for $i(t)$. One way to do this is to type the expression `i = DIF(#n,t)`, where **n** is the statement number of the expression for $q(t)$ (#44 for this example) or instead of #n just highlight the expression for $q(t)$ and move it into the appropriate place in the work area with the **F3** key. The function displays as

$$48: \ i = \frac{d}{dt}\left[\left[-\frac{11}{4}\right] \hat{e}^{-2t} + \frac{33}{13}\hat{e}^{-3t} + \frac{11 \text{COS}(2t)}{52} + \frac{55 \text{SIN}(2t)}{52}\right]$$

Simplify this to obtain the following expression for $i(t)$.

49: $$i = \frac{11\,\hat{e}^{-2t}}{2} - \frac{99\,\hat{e}^{-3t}}{13} + \frac{55\,\mathrm{COS}(2t)}{26} - \frac{11\,\mathrm{SIN}(2t)}{26}$$

Probably, the best way to check that this function for the current does not exceed a certain amperage rating is to plot this function for $t > 0$ with a proper scale to see the global behavior. Highlight the expression for $i(t)$ and issue the **Plot** command. Use the **Delete, All** command, if there are previous plots being drawn. Set x-scale $= 5$ and y-scale $= 2$ with the **Scale** menu command. **Move** the cross to $x = 20$ and $y = 0$ and execute the **Center** command. **Plot** the function to obtain the following plot screen.

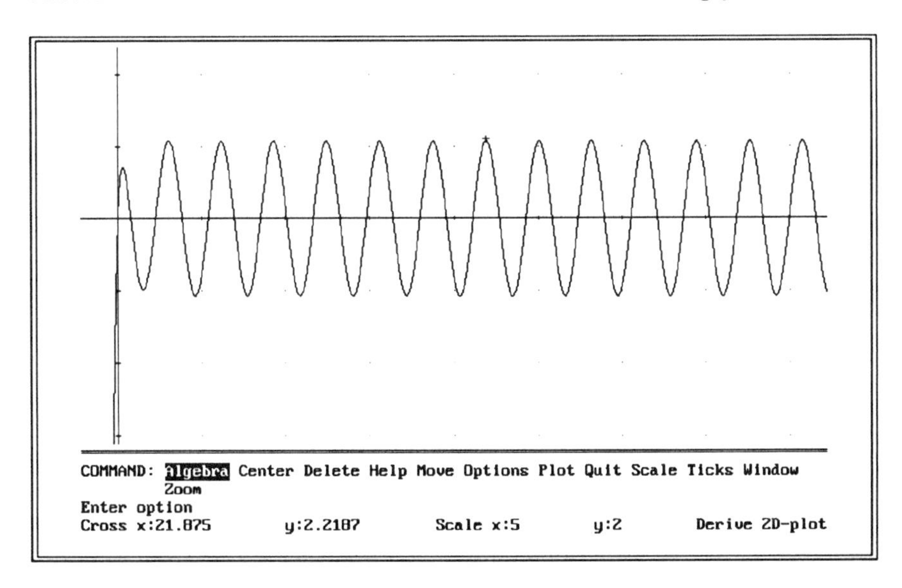

Move the cross using the direction arrows to determine that the maximum absolute current does not exceed a value of 2.3. Therefore, a 5 amp capacity (wiring and fuse) probably would be sufficient for this circuit.

While analyzing this solution, let's take note of the form of this long-term (steady-state) behavior. This behavior is more similar in form to the forcing function than the initial conditions. The initial conditions seem to affect only the short-term behavior. The graphing capabilities of Derive are very useful in making qualitative observations such as these.

Example 2.10: Differential System for Compartment Models

Teaching children to count is not as important as teaching them what counts.

— Anonymous

Subject: Solving a Differential System for Compartment Models in Biology Using Matrix Algebra

Problem: Biological organisms often are composed of component parts called compartments. Examples are the organs in the body—heart, lungs, stomach, and liver or the different parts of the blood—red blood cells, plasma, and white blood cells. Individual compartments may interact with one another and it's that interaction that is modeled by a system of differential equations. We want to find the concentration of a given substance in each compartment.

By assuming constant-volume and well-mixed compartments, mass transport equations for each compartment are derived. The differential model for a system with three compartments whose only exit to the outside environment is through the first compartment is written as

$$y_1' = -(k_{21} + k_{13} + k_{01})y_1 + k_{12}y_2 + k_{13}y_3 + b_1(t),$$

$$y_2' = k_{21}y_1 + -(k_{12} + k_{32})y_2 + k_{23}y_3 + b_2(t),$$

$$y_3' = k_{31}y_1 + k_{32}y_2 - (k_{13} + k_{23})y_3 + b_3(t),$$

where the k_{ij} are constant transfer coefficients that describe the ease of flow of the substance between compartments i and j. A 0 in the subscript signifies the interface with the external environment. A coefficient of 0 means no transfer of the substance and a coefficient of 1 means completely unimpeded flow. The b_i are direct input/output functions caused by the manufacture or use of the substance within the compartment.

If the following values for the transfer coefficients and input functions are given:

$k_{01} = 0.2$ $\quad k_{12} = 0.1$ $\quad k_{13} = 0.15$

$k_{21} = 0.3$ $\quad k_{32} = 0.15$ $\quad k_{23} = 0.25$

$k_{31} = 0.25$ $\quad b_2 = 0$ $\quad b_3 = 0.01$

$b_1 = at^2 + bt + c,$

then the equation in matrix-vector form becomes

$$\vec{y}' = \begin{bmatrix} -0.65 & 0.1 & 0.15 \\ 0.3 & -0.25 & 0.25 \\ 0.25 & 0.15 & -0.4 \end{bmatrix} \vec{y} + \begin{bmatrix} at^2 + bt + c \\ 0.0 \\ 0.01 \end{bmatrix}.$$

This is best solved using Derive with the steps presented in Section 2.8. In outline form these steps are: i) enter the matrix and find its eigenvalues with the commands **EIGENVALUES** and **Simplify**; ii) find the eigenvectors by solving three matrix equations (one for each eigenvalue) either using Derive or, if obvious, by hand; iii) form and enter into Derive the fundamental matrix, Φ; iv) **Author** the integral formula for the particular solution, $\vec{y}_p$, into Derive and **Simplify**; and v) use the result to **Author** the general solution, $\vec{y}_g$, and **Manage, Substitute** the initial values to get the solution. The results of using Derive to do these steps for this problem are shown below. Some statements are truncated.

```
                     ┌ -0.65   0.1   0.15 ┐
2:   EIGENVALUES     │  0.3  -0.25  0.25  │
                     └  0.25   0.15 -0.4  ┘

          ┌             ┌       ┌ 184 √8391 ┐          ┐                           ┌      ┌ 1
          │             │  ATAN │ ───────── │          │                           │ ATAN │ ─
          │             │       └   25173   ┘    5 π   │                           │      └
3:        │      √7 COS │ ───────────────────  - ───   │                    √7 COS │ ──────────
          │             └          3                6   ┘      13                  └
          │w = ──────────────────────────────────────────── - ──, w = ─────────────────────────
          └                        6                            30
```

```
      ┌┌ -0.65   0.1   0.15 ┐                             ┐
5:    ││  0.3  -0.25  0.25  │ - w IDENTITY_MATRIX (3)     │ · [a, b, c]
      └└  0.25   0.15 -0.4  ┘                             ┘

       ┌  a (20 w + 13)     b     3 c    3 a    b (4 w + 1)     c    a    3 b
6:     │- ───────────── + ──── + ────,  ──── - ─────────── + ──,  ── + ────
       └       20          10     20     10          4          4    4    20

       ┌    ┌    ┌            ┌       ┌ 184 √8391 ┐         ┐        ┐      ┐
       │    │    │            │  ATAN │ ───────── │         │        │      │
       │    │    │            │       └   25173   ┘    5 π  │        │      │
       │    │    │     √7 COS │ ──────────────────── - ───  │        │      │
7:     │    │    │            └          3               6   ┘   13   │      │
       │  a │ 20 │ ──────────────────────────────────────────── - ──  │ + 13 │
       │    └    └                       6                         30  ┘      ┘    b
       │- ──────────────────────────────────────────────────────────────────── + ── +
       └                                 20                                       10

8:    [a = 0, b = 0, c = 0]
```

Unfortunately, we are not able to complete the solution steps as given. Derive is unable to obtain the eigenvectors since it returns only the trivial solution, $\vec{0}$, as the solution to the matrix-vector equation. This is a *limitation* of Derive. More manipulations could be tried, but there is no fixed procedure to follow and it's usually best to use another, more appropriate or specialized, computational tool when this occurs. However, this one example does not mean that Derive cannot solve 3x3 systems of equations. Even though this kind of problem is very difficult, Derive usually does quite well on systems like this. It is worth trying to use Derive for these kinds of problems. Don't get frustrated or upset if the software cannot solve a difficult problem like this one. Think of all the problems it can help you with and understand that all software packages have limitations.

Example 2.11: Fourier Series

A mathematician, like a painter or a poet, is a maker of patterns. If his patterns are more permanent than theirs, it is because they are made with ideas.

— Godfrey Hardy [1940]

Subject: The Fourier Series

Problem: One of the commons tasks often needed to solve partial differential equations is to expand functions in Fourier series. Derive can help in this task in several ways. Derive's capabilities in this area are demonstrated in the following examples.

Solution: Derive has the command **FOURIER** in the utility file INTEGRAL.MTH that produces Fourier series approximations to functions. The utility file must be loaded into the work space through execution of the **Transfer, Load** command before the command can be used. The in-line command **FOURIER**$(u(x), x, a, b, n)$ produces the Fourier series approximation with n terms (both sine and cosine) to $u(x)$ from $x = a$ to $x = b$. The command evaluates the coefficients of the series through their integral definition. This display of a Derive screen shows the **FOURIER** command (truncated) and its in-line documentation as provided in the INTEGRAL.MTH file.

```
10:  "The Fourier series approximation of u(x) from x=a to x=b for n terms:"

                                          b                       ┌          ┌ 2 π k_ x
                                 1       ⌠              2      n  │          │ ────────
11:  FOURIER (u, x, a, b, n) := ─────── ⌡  u dx + ─────── Σ  │COS │
                               b - a     a           b - a  k_=1  │          └  b - a
                                                                  └
```

Just how many terms should be used to get a good approximation for $u(x)$? Let's try several different Fourier series approximations for $u(x) = x^3, -3 < x < 3$. To start with, let $n = 2, 4$, and 6. The sequence of input commands and outputs after executing **Simplify** are shown in the following Derive screen images.

```
AUTHOR expression: fourier(x^3,x,-3,3,2)

Enter expression
User                                   Free:93%              Derive Algebra
```

12: FOURIER $(x^3, x, -3, 3, 2)$

13: $\frac{54\,(\pi^2 - 6)\,\mathrm{SIN}\left[\frac{\pi x}{3}\right]}{\pi^3} - \frac{27\,(2\pi^2 - 3)\,\mathrm{SIN}\left[\frac{2\pi x}{3}\right]}{2\pi^3}$

16: FOURIER $(x^3, x, -3, 3, 6)$

17: $-\frac{3\,(6\pi^2 - 1)\,\mathrm{SIN}\,(2\pi x)}{2\pi^3} + \frac{54\,(25\pi^2 - 6)\,\mathrm{SIN}\left[\frac{5\pi x}{3}\right]}{125\pi^3}$

$-\frac{27\,(8\pi^2 - 3)\,\mathrm{SIN}\left[\frac{4\pi x}{3}\right]}{16\pi^3} + \frac{6\,(3\pi^2 - 2)\,\mathrm{SIN}\,(\pi x)}{\pi^3}$

$-\frac{27\,(2\pi^2 - 3)\,\mathrm{SIN}\left[\frac{2\pi x}{3}\right]}{2\pi^3} + \frac{54\,(\pi^2 - 6)\,\mathrm{SIN}\left[\frac{\pi x}{3}\right]}{\pi^3}$

```
             3
14:  FOURIER (x  , x, -3, 3, 4)

                 2             ┌ 4 π x ┐
        27 (8 π   - 3) SIN  │ ─────── │                2
                               └   3   ┘       6 (3 π  - 2) SIN (π x)
15:  - ────────────────────────────────── + ──────────────────────────
                          3                               3
                     16 π                                π
```

```
              2              ┌ 2 π x ┐             2              ┌ π x ┐
     27 (2 π   - 3) SIN  │ ─────── │      54 (π   - 6) SIN  │ ───── │
                             └   3   ┘                            └  3  ┘
  - ────────────────────────────────── + ──────────────────────────────
                     3                                3
                2 π                                  π
```

The additional lines in the output are the part of the expression hidden off the right-side of the screen. Of course, this part of an expression can accessed through careful use of the direction keys to move the highlight to the right through the expression. The direction keys are used to control the level of the highlight. The ↓ and ↑ change the level in the expression, and the → and ← move the highlight in the appropriate direction at the current level of the highlight.

In order to see the behavior as n changes, we need to plot these approximations and the original function. There are two ways to do this efficiently. One way is to put the four functions as separate components into a vector and plot the entire vector at once with the `Plot, Plot` command. The other way is to divide the screen into two windows and plot the functions one at a time. We'll show and explain the second way. Try the first way on your own.

First, execute the **`Window, Split`** command. Make the split **`Horizontal`** at line 15. this divides the screen into a large and small part. Then **`Window, Designate`** this larger window for **`2D-Plot`**. Now as **`Plot`** and **`Algebra`** commands are issued, Derive automatically will move back and forth from the plotting window and algebra window. Another way to move back and forth between the two windows is with the **`Windows, Next`** or **`Windows, Previous`** commands. Derive indicates the active window by background shading around the number label of the active window.

Move to the algebra window (**Algebra** or **Window, Next**) and use the direction keys (↑ or ↓) to highlight the first function ($n = 2$). Move back to the plotting window and plot this function through the **Plot, Plot** command. The result is the following display.

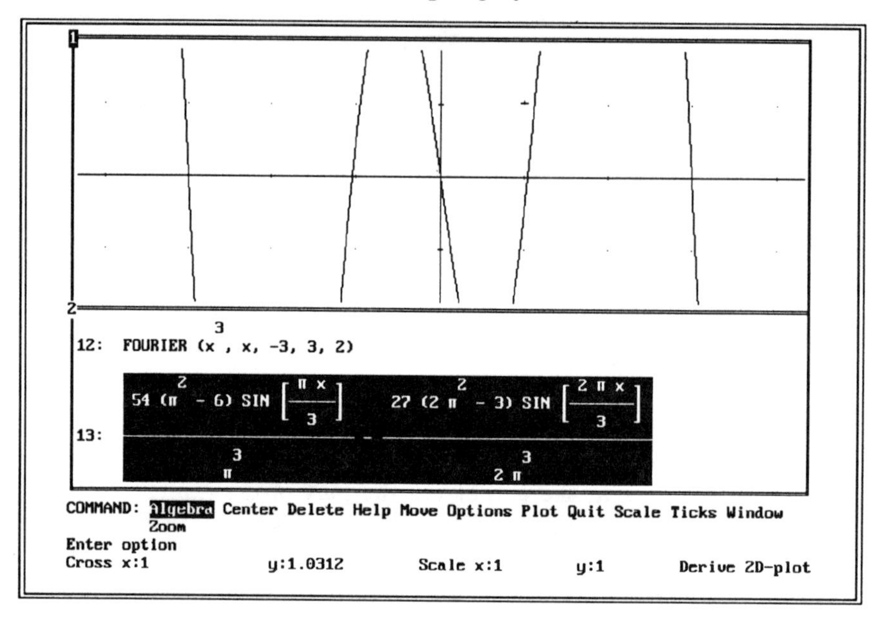

Not enough of the function is shown to determine its global behavior. Try changing the scale using the **Scale** command. Set the x-scale to 3 and the y-scale to 15. Now, reissue the **Plot** command. This plot is much better. It shows the periodic behavior of the function and its maximum and minimum values.

Now, it is only a matter of moving back and forth between the two windows (using **Plot** and **Algebra**), highlighting the next function and plotting it on the same axes as the previous plots. On a color monitor, each plot is done in a different color. The resulting black-and-white plot of all four functions is as shown.

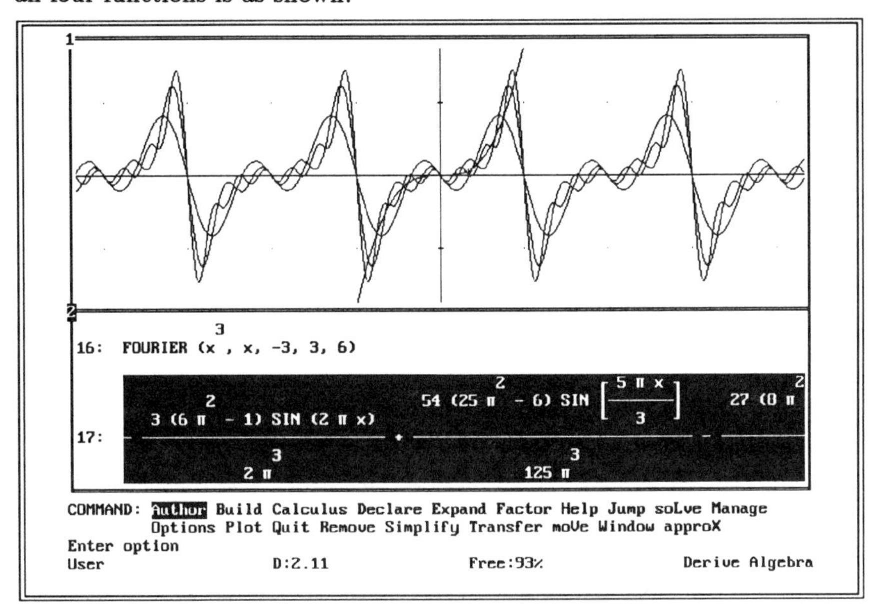

As the plotting takes place, notice the improvement in the series approximations as n increases. But at the same time the oscillations don't disappear. This is a phenomenon discussed in many differential equation textbooks. Once this plot window is no longer needed, close the plotting window by executing the **Window, Close** command.

Let's try to find the Fourier series approximation to another function, which will result in some tougher integration. If $u(x) = \sin(x^2)$, the command

```
FOURIER(sin(x^2),x,0,2,4)
```

sets up the integration for the series approximation to four terms of $u(x)$ in the region $0 < x < 2$. **Simplify** performs the integration. The results are displayed in the following figure.

18: FOURIER $(\mathrm{SIN}\,(x^2),\ x,\ 0,\ 2,\ 4)$

19: $\int_0^2 \mathrm{SIN}\,(x^2)\ \mathrm{COS}\,(4\ \pi\ x)\ dx\ \mathrm{COS}\,(4\ \pi\ x) + \int_0^2 \mathrm{SIN}\,(x^2)\ \mathrm{SIN}\,(4\ \pi\ x)\ dx\ \mathrm{SIN}\,($

Since integral set-ups remain in the simplified expression, this means Derive is unable to perform the integration in the **Exact** mode. Change the mode to **Mixed** or **Approximate** through the **Options, Precision** menu commands. Now reexecute the **Simplify** command on the **FOURIER** in-line command. Do this by moving the highlight to that expression or entering the appropriate expression number when displayed in the **Simplify** command.

This time numerical quadrature is used to perform the definite integration. The system will take quite a long time and may beep and issue the warning "**dubious accuracy**". This means the numerical integration algorithm is having difficulties satisfying the error tolerance set in the digits mode of the **Precision** command. Let the algorithm continue its work. The display with the result extended on extra lines is shown.

20: $- \frac{303}{15643}\ \mathrm{COS}\left[\frac{1420}{113}\ x\right] + \frac{4065}{62998}\ \mathrm{SIN}\left[\frac{1420}{113}\ x\right] - \frac{510}{12961}\ \mathrm{COS}\left[\frac{1065}{113}\ x\right]$

$+ \frac{714}{7897}\ \mathrm{SIN}\left[\frac{1065}{113}\ x\right] - \frac{2159}{17338}\ \mathrm{COS}\left[\frac{710}{113}\ x\right] + \frac{8777}{62967}\ \mathrm{SIN}\left[\frac{710}{113}\ x\right]$

$\frac{4483}{8355}\ \mathrm{COS}\left[\frac{355}{113}\ x\right] - \frac{79}{385}\ \mathrm{SIN}\left[\frac{355}{113}\ x\right] + \frac{8933}{22200}$

Now, in light of the dubious accuracy, we should compare the plot of the approximation with that of the actual $u(x)$ function. Unfortunately there is no way to do this efficiently. If the expressions are put in a vector as the two components, Derive thinks the expression is in polar form. So first **Plot** $\sin(x^2)$, then return to the **Algebra** window and highlight the Fourier approximation. Issue the **Plot, Plot** command to plot the second function after the first plot is redone. These plots using the default scale parameters are as follows.

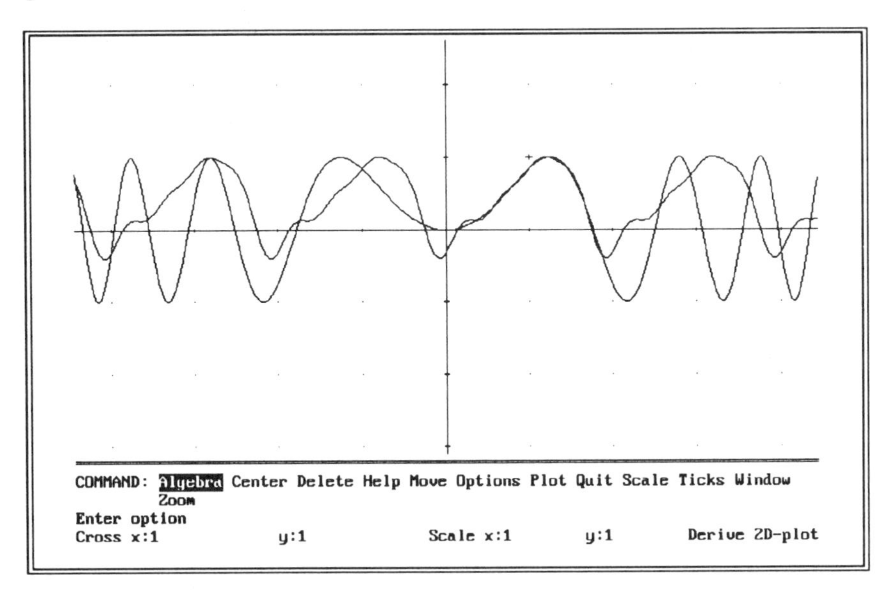

Even though the plot of the Fourier approximation looks strange, it matches the function quite well in the specified interval, $0 < x < 2$. The graph exhibits the periodic property of the Fourier series on the rest of the plot interval.

The utility file doesn't have commands for the Fourier sine series or the Fourier cosine series. If these expansions are needed, the actual definition for the coefficients for these series could be entered through the **Author** command. Probably, the easiest way to input the definitions for these two series is to modify the definition of the existing **FOURIER** command using the direction arrows, the highlight, the function keys, and the special control keys discussed in Section 1.2. For instance, a double sine series for $f(x, y)$ involves the following integration to determine the coefficients

$$A_{mn} = \frac{4}{bc} \int_0^c \int_0^b f(x, y) \sin\left(\frac{m\pi}{b} x\right) \sin\left(\frac{n\pi}{c} y\right) dx dy.$$

If $f(x, y) = xy, 0 < x < 1, 0 < y < 1$, then the Derive in-line command for this integration is

`INT(INT(xy sin(m pi x/b) sin(n pi y/c),x,0,b),y,0,c)`.

When the **Precision** is still in the **Approximate** or **Mixed** modes, this operation is performed through the **Author** and **Simplify** commands to obtain the following display with 3 digits of accuracy. Once again, the additional line under the expression for the approximation is the rest of the expression usually hidden off the right-side of the screen.

$$21: \int_0^c \int_0^b x\, y\, \mathrm{SIN}\left[\frac{m\,\pi\,x}{b}\right] \mathrm{SIN}\left[\frac{n\,\pi\,y}{c}\right] dx\, dy$$

$$23: \left[\frac{0.101\, b^2\, c^2\, \mathrm{COS}\,(3.14\, n)}{m\, n} - \frac{0.0322\, b^2\, c^2\, \mathrm{SIN}\,(3.14\, n)}{m\, n^2}\right] \mathrm{COS}\,(3.14\, m)$$

$$+ \left[\frac{0.0102\, b^2\, c^2\, \mathrm{SIN}\,(3.14\, n)}{m^2\, n^2} - \frac{0.0322\, b^2\, c^2\, \mathrm{COS}\,(3.14\, n)}{m^2\, n}\right] \mathrm{SIN}\,(3.14\, m)$$

Example 2.12: Laplace Transform

Curiouser and curiouser!

—Lewis Carroll *Alice in Wonderland* [1856]

Subject: Using Laplace Transforms to Solve Differential Equation with Discontinuous Forcing Functions

Problem: Find the solution to the initial-value problem

$$O(y) := y'' + 4y = u(t) = \begin{cases} 4t & \text{if } 0 \le t < 1 \\ 4 & \text{if } t \ge 1 \end{cases}$$

with $y(0) = 1$ and $y'(0) = 0$.

Solution: First use the **STEP** function to define the forcing function $u(t)$ in this problem. This is done with the **Declare, Function** command with function name input as u and definition as `STEP(a-t)b + STEP(t-a)c`. This produces the Derive function

$$u(a, b, c, t) = \begin{cases} b & \text{if } t < a \\ c & \text{if } t > a \end{cases}$$

Therefore, the forcing function for this problem is entered as `u(1,4t,4,t)` and the following display results.

```
1:   U (a, b, c, t) := STEP (a - t) b + STEP (t - a) c

2:   U (1, 4 t, 4, t)
```

The plot of this function using **Plot, Zoom, Both, Out, Plot** reveals its stepped behavior.

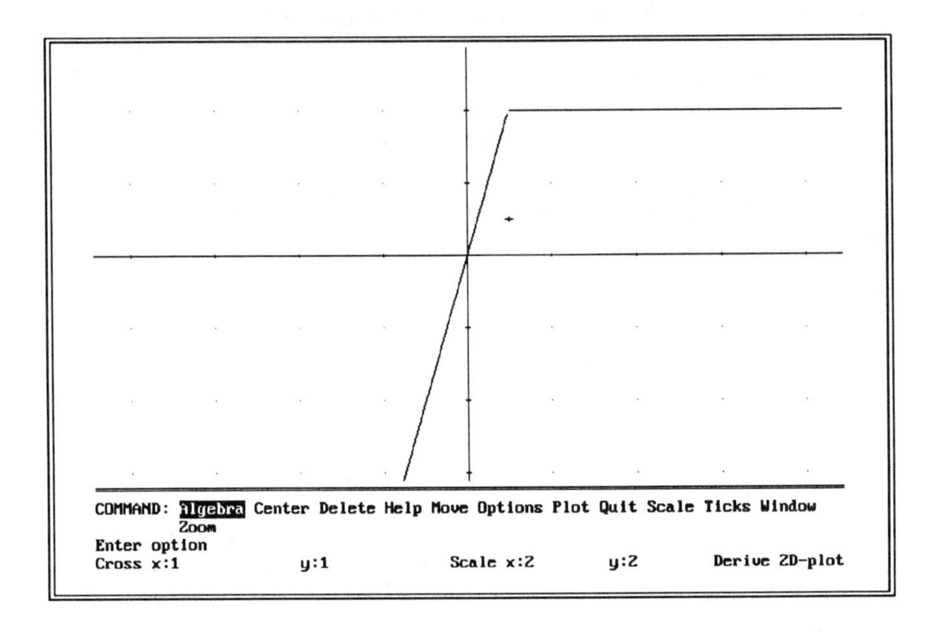

To find the Laplace transform $L(u(t))$, the formula

$$L(u(t)) = \int_0^\infty e^{-st} u(t) dt, \quad s > 0,$$

can be used. To implement this with Derive, first declare s positive with the **Declare, Variable** command. Then the integrand is formed using **Author** command and the following input.

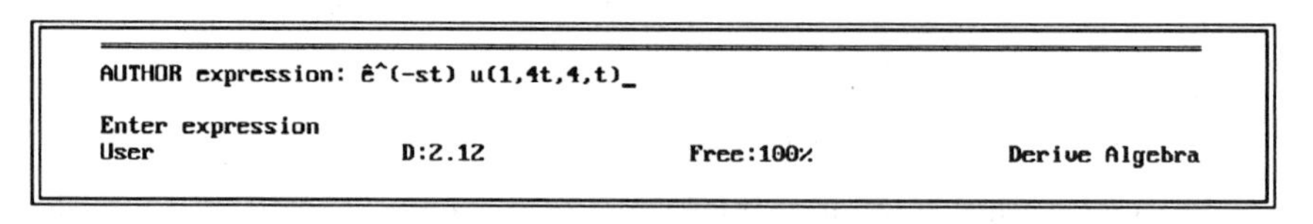

The menu command **Calculus, Integrate** is used to obtain the transform. The user is prompted for the variable of integration: t; the lower limit: 0; and the upper limit: inf. **Simplify** and the result is as shown.

```
3:  ê^(- s t) U (1, 4 t, 4, t)

      ∞
4:  ∫  ê^(- s t) U (1, 4 t, 4, t) dt
      0

     4      4 ê^(-s)
5:  ---  -  -------
     s^2      s^2
```

This integral function is already defined in the utility file INTEGRAL.MTH as **LAPLACE**(u, s, t). However, sometimes it is not worth the hassle to load a utility file to perform one simple integration. If several transforms or other functions in the INTEGRAL file are needed, then we can load the utility file and use its functions.

By taking the Laplace transform of the left side of the equation by hand, the transformed equation becomes

$$s^2 Y(s) - y'(0)s - y(0) + 4Y(s) = \frac{4 - 4e^{-s}}{s^2}.$$

Author this equation with the initial conditions substituted for $y'(0)$ and $y(0)$ by entering

```
AUTHOR expression: s^2Y-1+4Y=4 / s^2 - 4 ê^(-s) / s^2_

Enter expression
Simp(4)              D:2.12              Free:100%              Derive Algebra
```

Then, **soLve** for $Y(s)$ to obtain the following screen image.

```
               ê^(-s) ((s^2 + 4) ê^s - 4)
7:    y  =  ------------------------------
                   s^2 (s^2 + 4)
```

In order to find the inverse Laplace transform, use the **Expand** command to get the partial fraction decomposition form of the forcing function on the right-hand side of the equation as shown. Just press **Enter** or **s Enter** when queried for the **EXPAND variable**.

```
           -s        -s
           ê         ê        1
8:   y = ─────── - ───── + ───
           2         2        2
          s  + 4    s        s
```

Unfortunately, Derive does not provide a direct way to find the inverse transform. Derive does not do the contour integration which is necessary to find inverse Laplace transforms by means of the definition. Neither does Derive keep a table of values for this operation, so it must be done by hand from tables. This is not difficult for this problem once the partial fraction decomposition is performed. The solution is

$$y(t) = 1/2\sin(2t-2)\mathrm{STEP}(t-1)-(t-1)\mathrm{STEP}(t-1)+\cos(2t)-1/2\sin(2t)+t.$$

Before plotting the solution, let's check to make sure it solves the equation and initial conditions. First, define the differential operator $O(y)$ by

```
AUTHOR expression: o(y):=dif(y,t,2) +4y _

Enter expression
Expd(7)               D:2.12               Free:100%               Derive Algebra
```

Then, **Author** the solution by

```
AUTHOR expression: y:=.5sin(2t-2)step(t-1)-(t-1)step(t-1)+cos(2t) -.5sin(2t)+t

Enter expression
User                  D:2.12               Free:100%               Derive Algebra
```

Enter $O(y)$ and **Simplify** to obtain the value of the operator as shown in the following figure.

```
9:   D (y) := [d/dt]^2 y + 4 y

10:  y := 0.5 SIN (2 t - 2) STEP (t - 1) - (t - 1) STEP (t - 1) + COS (2 t) - 0.

11:  D (y)

12:  2 (t + 1) - 2 |t - 1|
```

Compare this expression with the right-side of the equation $u(t)$ by *Authoring* their difference and executing **Simplify**. Just as we hoped, the result indicates the solution is correct.

```
13:  2 (t + 1) - 2 |t - 1| - U (1, 4 t, 4, t)

14:  0
```

Check the initial condition $y(0) = 1$ by **Manage, Substitute** 0 for t in the expression for the solution. This result further verifies the solution.

```
16:  0.5 SIN (2 t - 2) STEP (t - 1) - (t - 1) STEP (t - 1) + COS (2 t) - 0.5 SIN

17:  0.5 SIN (2 0 - 2) STEP (0 - 1) - (0 - 1) STEP (0 - 1) + COS (2 0) - 0.5 SIN

18:  1
```

Check the initial condition $y'(0) = 0$ by highlighting the solution and executing **Calculus, Differentiate** with respect to t and **Simplify**. Use the **Manage, Substitute** command to substitute 0 for t and **Simplify**. All these operations are shown in the following figure.

19: $\frac{d}{dt}$ (0.5 SIN (2 t - 2) STEP (t - 1) - (t - 1) STEP (t - 1) + COS (2 t) - 0.5

20: $\left[\frac{\text{COS (2 t - 2)}}{2} - \frac{1}{2}\right]$ SIGN (t - 1) + $\frac{\text{COS (2 t - 2)}}{2}$ - COS (2 t) - 2 SIN

21: $\left[\frac{\text{COS (2 0 - 2)}}{2} - \frac{1}{2}\right]$ SIGN (0 - 1) + $\frac{\text{COS (2 0 - 2)}}{2}$ - COS (2 0) - 2 SIN

22: 0

Now, the solution can be plotted. First, highlight the solution and then execute the **Plot, Plot** command. Make sure to execute **Delete, All** in order to delete any previous plots. This graph gives a nice visualization of the local behavior of the solution near the origin.

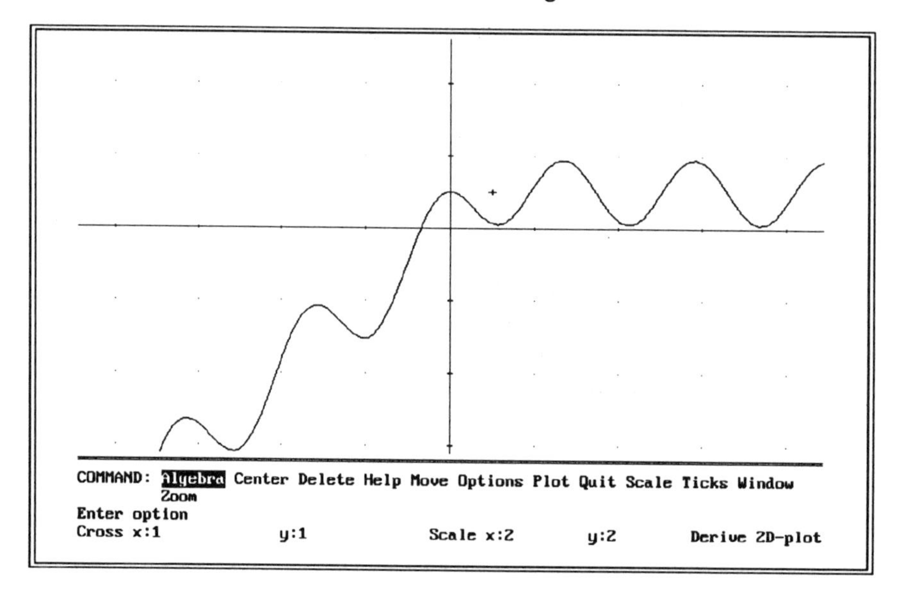

In order to obtain an plot with more global behavior, execute **Zoom, Both, Out** twice. The resulting plot is as follows:

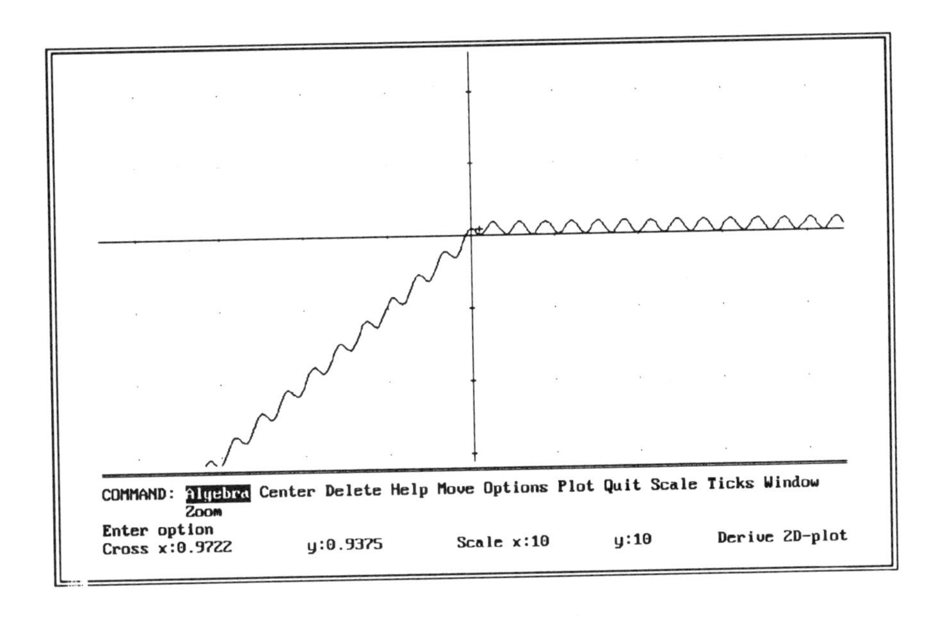
COMMAND: Algebra Center Delete Help Move Options Plot Quit Scale Ticks Window
Zoom
Enter option
Cross x:0.9722 y:0.9375 Scale x:10 y:10 Derive 2D-plot

Example 2.13: Separation of Variables

Education is not received. It is achieved.

— Anonymous

Subject: Separation of Variables for the Wave Equation

Problem: The vertical displacement $u(x,t)$ of a vibrating string of length l is determined from solving the wave equation, a hyperbolic partial differential equation of the form

$$k^2 \frac{\partial^2 u}{\partial x^2} = \frac{\partial^2 u}{\partial t^2}, \quad 0 < x < l, \quad t > 0,$$
$$u(0,t) = 0, \quad u(l,t) = 0,$$
$$u(x,0) = bx(l-x), \quad \frac{\partial u(x,0)}{\partial t} = c.$$

Here, x is the horizontal location, t is time, k is a parameter relating string tension and mass, b is a parameter controlling the amplitude of the initial displacement, and c is a parameter controlling the initial velocity of the string. These initial conditions simulate plucking the string at its middle (at $x = l/2$).

Solve this equation and investigate its solution and standing waves at $t = 1$ for specified values of the parameters k, b, and c.

Solution: First, let's use Derive's plotting capability to visualize the initial geometry of the problem. **Author** the initial displacement into the work area by entering

```
AUTHOR expression: bx(l-x)

Enter expression
                                    Free:100%            Derive Algebra
```

Derive's 2-D plotter won't plot an expression with extra unknown parameters such as b and l because they make it a higher-dimensional expression, so substitute a value of 1 for b and 3 for l using the **Manage, Substitute** command. Just hit 1 for the b value, 3 for the l value, and **Enter** for the x value. Now, execute the **Plot, Plot** command. The plot of the initial displacement of the string with these parameter values and using the default plotting parameters values is as shown.

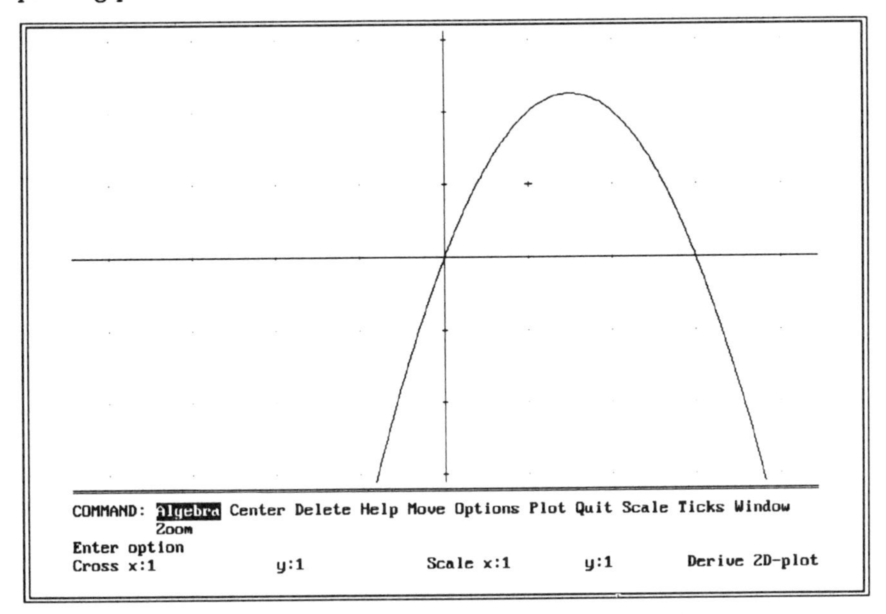

Derive plots the function over the entire domain of the x-axis shown in the window, even if a specific interval is established for x using the **Declare, Variable** and **Interval** menu commands. The actual function describing the initial condition exists only between $x = 0$ and $x = l = 3$.

The solution technique of separation of variables assumes that the solution $u(x,t)$ is a product of a function of x only $(X(x))$ and a function of t only $(T(t))$, so that $u(x,t) = X(x)T(t)$. Making this substitution, the equation becomes

$$k^2X''T = XT''.$$

Separate the variables by dividing common terms and equating to a constant, w, to get

$$\frac{X''}{X} = \frac{T''}{k^2T} = w.$$

We use w instead of the more common λ because λ is one of the Greek characters that Derive does not have available for use. This leads to the need to solve the two ordinary differential equations

$$X'' - wX = 0$$

and

$$T'' - wk^2T = 0.$$

The boundary conditions for the first equation are $X(0) = 0$ and $X(l) = 0$.

Commands from Derive's utility file ODE2.MTH can be used to solve for $X(x)$ and $T(t)$ for the three possibilities for w, $w > 0$, $w < 0$, and $w = 0$. The utility file is loaded into the work area with the **Transfer, Load** or **Transfer, Merge** commands.

For the situation where $w > 0$, the command **Declare, Variable** can be used to declare w **Positive**, but why bother. We'll keep track of the sign of w as we go through the three cases. Now solving for X, **Author** and **Simplify** the in-line command, `LIN2_RED_CCF_DISC(0,-w)`. The result is

```
37:  LIN2_RED_CCF_DISC (0, -w)

38:  4 w
```

Since the resulting value for the discriminant is positive, the command `LIN2_RED_CCF_POS(0,-w,x)` is used to obtain the solution upon executing **Simplify** as shown.

```
39:  LIN2_RED_CCF_POS (0, -w, x)

           √w x        - √w x
40:  @1 ê       + @2 ê
```

The @1 and @2 are arbitrary constants.

The **IMPOSE_BV2** command does not help find the eigenvalues for w, since for these conditions Derive always provides the obvious solution of @1 and @2 equal to 0. However, in this case the boundary conditions of $@1 + @2 = 0$ and $@1e^{\sqrt{w}l} + @2e^{-\sqrt{w}l} = 0$ do lead to $@1 = 0$ and $@2 = 0$.

The easiest way to change the sign on w is to change the first equation to $X'' + wX = 0$. This changes the argument in the discriminant function to w and changes the solution function to **LIN2_RED_CCF_NEG**. For this new equation, the screen output of the solution operations is shown.

```
42:  LIN2_RED_CCF_NEG (0, w, x)

43:  @2 COS (√w x) + @1 SIN (√w x)
```

Once again, the command **IMPOSE_BV2** is not appropriate. Derive doesn't have a way to restrict variables, like n, to integer values. The two boundary conditions lead to @2 = 0 and $\sin\sqrt{w}l = 0$. Therefore, $\sqrt{w} = n\pi/l$, $n = 1, 2, \ldots$. For this case, the equation for $T(t)$ becomes

$$T'' + \frac{n^2\pi^2}{l^2}k^2T = 0.$$

The commands to solve this equation for T are similar to those for the equation in x. The screen display of these commands is provided.

```
                               [       2  2  2   ]
                               [      n  π  k    ]
44:  LIN2_RED_CCF_DISC         [0, ------------  ]
                               [         2       ]
                               [        l        ]

           2  2  2
        4 k  n  π
45:  - ------------
             2
            l

                              [       2  2  2        ]
                              [      n  π  k         ]
46:  LIN2_RED_CCF_NEG         [0, ------------, t    ]
                              [         2            ]
                              [        l             ]

47:  @1 SIN [k n π t / l] SIGN (k l n) + @2 COS [k n π t / l]
```

For the last case of $w = 0$, the equation is simply $X'' = 0$ which has the solution $@1 + @2x$. The command `LIN2_RED_CCF_0(0,x)` could have been used to get this result. However, the boundary conditions force the trivial solution for this case, also.

Therefore, the solutions for $u(x,t)$ have the form

$$\left[(a(n)\cos\frac{n\pi k}{l}t + b(n)\sin\frac{n\pi k}{l}t)\right]\sin\frac{n\pi}{l}x.$$

By the superposition principle,

$$u(x,t) = \sum_{n=1}^{\infty}\left[(a(n)\cos\frac{n\pi k}{l}t + b(n)\sin\frac{n\pi k}{l}t)\right]\sin\frac{n\pi}{l}x.$$

For this solution, the initial conditions are

$$u(x,0) = \sum_{n=1}^{\infty} a(n)\sin\frac{n\pi}{l}x = bx(l-x).$$

and

$$\frac{\partial u(x,0)}{\partial t} = \sum_{n=1}^{\infty} b(n) \frac{n\pi k}{l} \sin \frac{n\pi}{l} x = c.$$

The coefficients, $a(n)$ and $b(n)$, are evaluated as Fourier coefficients by

$$a(n) = \int_0^l \frac{2}{l} bx(l-x) \sin \frac{n\pi}{l} x \, dx$$

and

$$b(n) = \int_0^l \frac{2}{n\pi k} c \sin \frac{n\pi}{l} x \, dx.$$

To evaluate these integrals, **Author** the expression of the integrand `2/l(bx(l-x)sin(n pi x/l))` and select the **Calculus, Integrate** menu command. Input the lower and upper limits, 0 and l, and **Simplify**. (Remember, the **Tab** key moves the cursor through the various parts of the menu.) The resulting display and expression for finding $a(n)$ are as follows.

$$48: \frac{2}{l}\left[b\,x\,(l-x)\,\mathrm{SIN}\left[\frac{n\,\pi\,x}{l}\right]\right]$$

$$49: \int_0^l \frac{2}{l}\left[b\,x\,(l-x)\,\mathrm{SIN}\left[\frac{n\,\pi\,x}{l}\right]\right] dx$$

$$50: -\frac{4\,b\,l^2\,\mathrm{COS}\,(n\,\pi)}{n^3\,\pi^3} - \frac{2\,b\,l^2\,\mathrm{SIN}\,(n\,\pi)}{n^2\,\pi^2} + \frac{4\,b\,l^2}{n^3\,\pi^3}$$

An alternate procedure for integration is to use the in-line command **INT**. The command for $b(n)$ is `INT(2c/(n pi k) sin(n pi x/l),x,0,l)`. Be careful to notice the difference between the l and 1 in the input string. Select the command **Simplify** to perform the operation. The following screen display shows the results of these operations.

```
                1
               ⌠     2 c       ┌ n π x ┐
         51:   ⌡   ─────── SIN │───────│ dx
                    n π k      └   1   ┘
                0

                 2 c l       2 c l COS (n π)
             ──────────── - ────────────────
         52:      2  2            2  2
               k n  π          k n  π
```

Since Derive still doesn't know n is an integer, it doesn't simplify $\sin(n\pi)$ to 0 or $\cos(n\pi)$ to $(-1)^{n-1}$. This is one of Derive's limitations. So with this type of manual simplification, the solution becomes

$$u(x,t) = \sum_{n=1}^{\infty} \left(\frac{4l^2 b}{\pi^3 n^3}(1-(-1)^n) \cos \frac{n\pi k}{l} t + \frac{2cl}{n^2 k \pi^2}(1+(-1)^n) \sin \frac{n\pi k}{l} t \right) \sin \frac{n\pi}{l} x.$$

Let's try plotting the first three approximate solutions ($m = 1, 2, 3$), where m is the upper limit of the index n in the summation, at $t = 1$ for the following specified values of the parameters: $b = 1.1, c = 1.5, k = 4.0, l = 3$. **Author** the general expression for the solution by entering the following input strings

```
AUTHOR expression: 4l^2b/(pi^3n^3)(1-(-1)^n)cos(npikt/l)_

Syntax error detected at cursor
Simp(51)              D:2.13               Free:100% Insert
```

```
AUTHOR expression: 2cl/(n^2k pi^2)(1+(-1)^n) sin(n pi kt/l)_

Enter expression
User                  D:2.13               Free:100% Insert
```

$$53: \quad \frac{4\, l^2\, b}{\pi^3\, n^3} \left(1 - (-1)^n\right) \text{COS}\left[\frac{n\, \pi\, k\, t}{l}\right]$$

$$54: \quad \frac{2\, c\, l}{n^2\, k\, \pi^2} \left(1 + (-1)^n\right) \text{SIN}\left[\frac{n\, \pi\, k\, t}{l}\right]$$

AUTHOR expression: sum((#53+#54)sin(n pi x/l),n,1,m)_

Enter expression
Simp(51) D:2.13 Free:100%

$$55: \quad \sum_{n=1}^{m} \left[\frac{4\, l^2\, b}{\pi^3\, n^3} \left(1 - (-1)^n\right) \text{COS}\left[\frac{n\, \pi\, k\, t}{l}\right] + \frac{2\, c\, l}{n^2\, k\, \pi^2} \left(1 + (-1)^n\right) \text{SIN}\left[\frac{n\, \pi}{\ }\right.\right.$$

Then **Manage, Substitute** these variable and parameter values. For $m = 1$, the expression, its truncated result, and its plot are shown. Remember, we are only interested in the region $0 < x < 3$. The resulting summation expression is shown. The additional lines are the continuation of the expression which are off the right of the screen in the Derive display.

$$57: \quad \sum_{n=1}^{1} \left[\frac{4\, 3^2\, 1.1}{\pi^3\, n^3} \left(1 - (-1)^n\right) \text{COS}\left[\frac{n\, \pi\, 4\, 1}{3}\right] + \right.$$

$$\left. + \frac{2\, 1.5\, 3}{n^2\, 4\, \pi^2} \left(1 + (-1)^n\right) \text{SIN}\left[\frac{n\, \pi\, 4\, 1}{3}\right]\right]$$

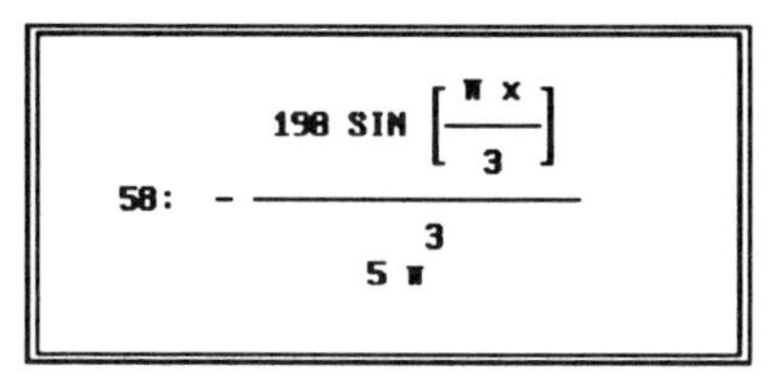

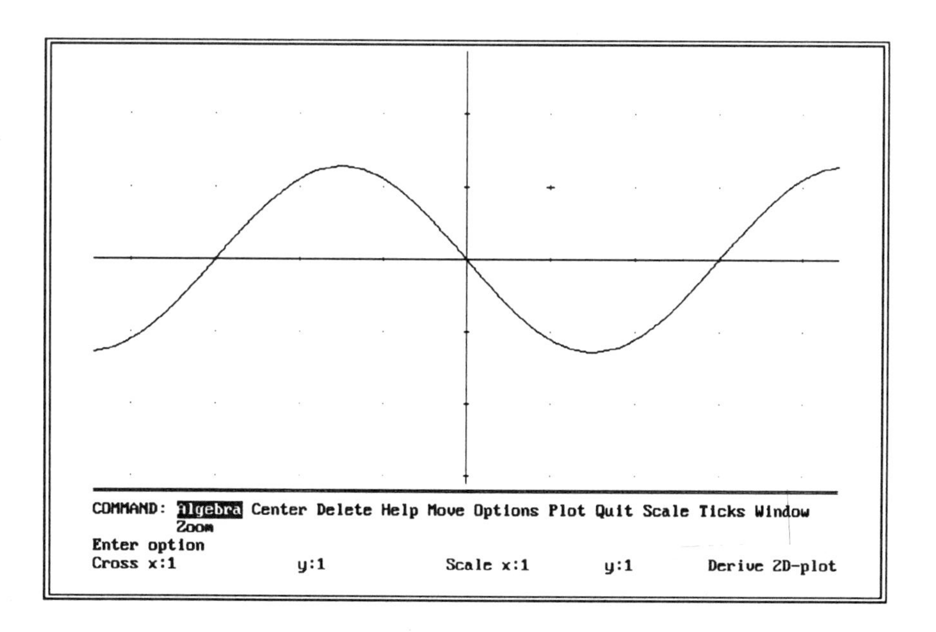

Before, producing and plotting the other two approximations ($m = 2$ and $m = 3$), let's create a better plotting region. Use the **Scale** command to change the plot scale for x (horizontal) to 0.5 and y (vertical) to 0.5 per tick mark. **Move** the cross to $x = 1.5$ and $y = -0.5$ and execute **Center**. The new approximation for $m = 2$ (truncated) and its plot (shown with the previous plot for $m = 1$) are as follows.

$$59: \quad \sum_{n=1}^{2}\left[\frac{4\,3^{2}\,1.1}{\pi^{3}\,n^{3}}\,(1-(-1)^{n})\,\cos\left[\frac{n\,\pi\,4\,1}{3}\right] + \frac{2\,1.5\,3}{n^{2}\,4\,\pi^{2}}\,(1+(-1)^{n})\,\sin\left[\frac{n}{}\right.\right.$$

$$60: \quad \frac{9\sqrt{3}\,\sin\left[\frac{2\,\pi\,x}{3}\right]}{16\,\pi^{2}} - \frac{198\,\sin\left[\frac{\pi\,x}{3}\right]}{5\,\pi^{3}}$$

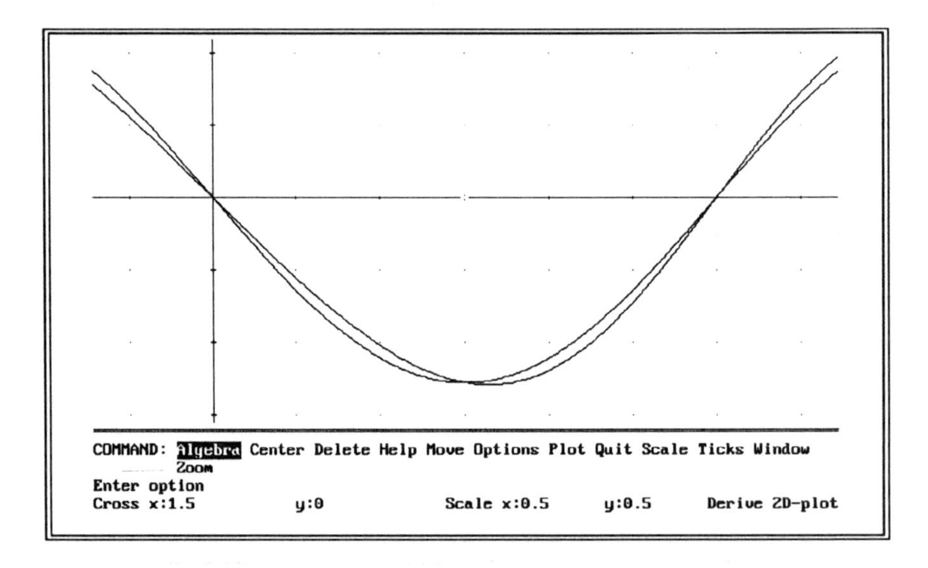

Now in a similar manner, return to the **Algebra** window to find the approximation with $m = 3$ and execute **Plot** to get all three plots drawn on the same axes.

61: $\sum_{n=1}^{3}\left[\frac{4\ 3^2\ 1.1}{\pi^3\ n^3}\left(1-(-1)^n\right)\cos\left[\frac{n\ \pi\ 4\ 1}{3}\right]+\frac{2\ 1.5\ 3}{n^2\ 4\ \pi^2}\left(1+(-1)^n\right)\sin\left[\frac{n}{}\right.\right.$

62: $$\frac{44\sin(\pi x)}{15\ \pi^3}+\frac{9\sqrt{3}\sin\left[\frac{2\ \pi\ x}{3}\right]}{16\ \pi^2}-\frac{198\sin\left[\frac{\pi\ x}{3}\right]}{5\ \pi^3}$$

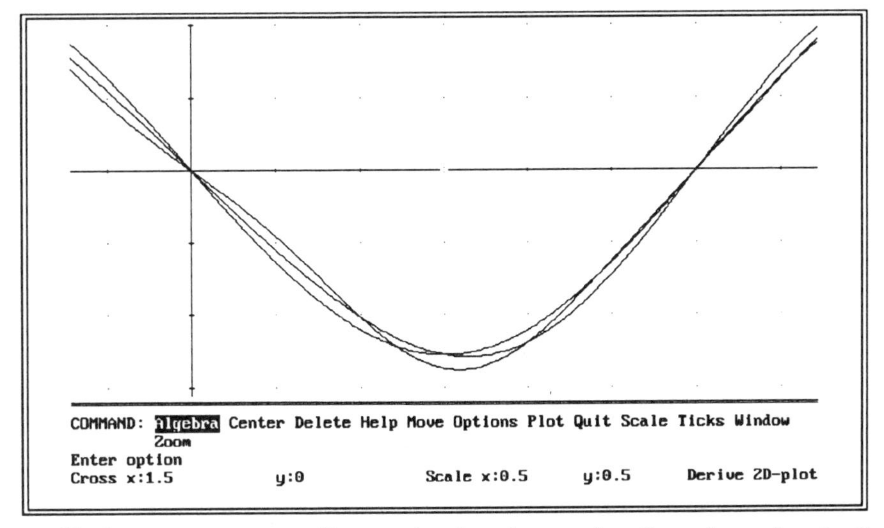

Various parameter studies can be done by varying the values of each of the parameters (b, c, k, and l). An example of this might be to determine the value of c which restricts the amplitude to values less than 0.2. Go ahead and try some of these for this problem by substituting different values for these parameters and plotting the resulting solutions. Similar parameter experiments are required in the several of the exercises in Chapter 3.

Now, let's look at a 4-term approximation to the series solution at various values of t ($t = 1, 2, 3$, and 4). Let's keep the previously specified values of the parameters: $b = 1.1, c = 1.5, k = 4.0, l = 3$. Using the general expression for the solution, **Manage, Substitute** these parameter values (with $m = 4$ and keeping x and t as unknown variables for now). **Simplify** this expression to obtain the 4-term approximation to $u(x,t)$. This expression is shown with the extra terms usually off the screen to the right wrapped around below the beginning of the expression.

$$64: \frac{9 \sin\left[\frac{16 \pi t}{3}\right] \sin\left[\frac{4 \pi x}{3}\right]}{32 \pi^2} + \frac{44 \cos(4 \pi t) \sin(\pi x)}{15 \pi^3}$$

$$+ \frac{9 \sin\left[\frac{8 \pi t}{3}\right] \sin\left[\frac{2 \pi x}{3}\right]}{8 \pi^2} + \frac{396 \cos\left[\frac{4 \pi t}{3}\right] \sin\left[\frac{\pi x}{3}\right]}{5 \pi^3}$$

Use the **Manage, Substitute** command to input the four values of t, one at a time, into this function. **Simplify** each function and put them together as components of a vector. The easiest way to do this is to **Author** the expression **[#a,#b,#c,#d]** where a, b, c, and d are the Derive statement numbers in the work area of the four functions. Plot all four functions at once on the same axes using the **Plot, Plot** menu command. The resulting plots are as shown. Only three different plots are shown since it turns out that the solutions for $t = 1$ and $t = 4$ are the same because the solution is periodic in t.

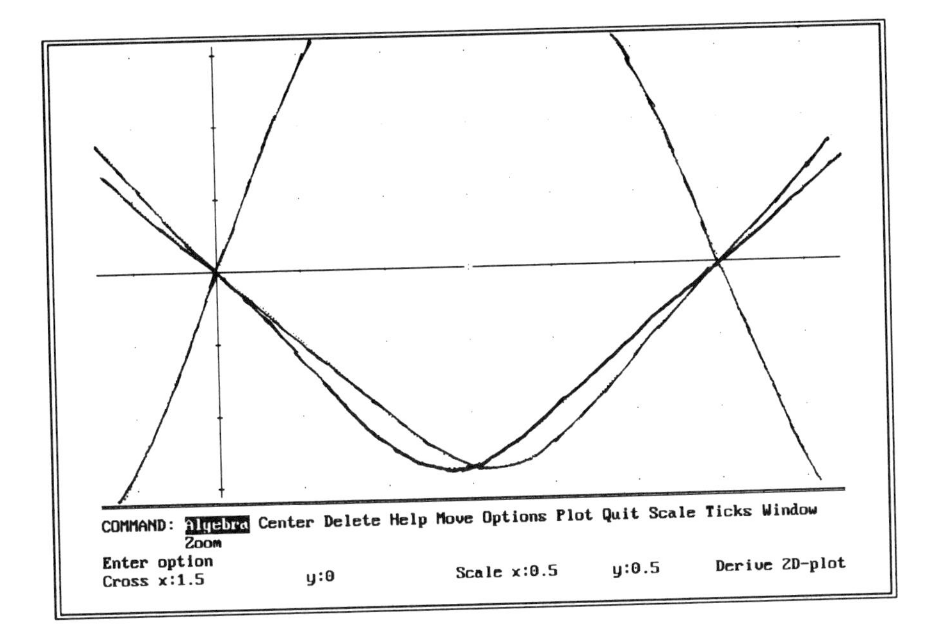
COMMAND: Algebra Center Delete Help Move Options Plot Quit Scale Ticks Window
Zoom
Enter option
Cross x:1.5
y:0
Scale x:0.5
y:0.5
Derive 2D-plot

Example 2.14: Bessel Equation

The most beautiful thing we can experience is the mysterious.

— Albert Einstein [1930]

Subject: Using Bessel Functions to Solve Bessel Equations

Problem: Two frequently occurring differential equations in science and engineering are the Bessel equation of order n and the parametric Bessel equation of order n. The two equations are given respectively by

$$x^2y'' + xy' + (x^2 - n^2)y = 0$$

and

$$x^2y'' + xy' + (\lambda^2x^2 - n^2)y = 0.$$

Some of the best-known applications of these equations are the determination of oscillations of a hanging chain or string, the theory of vibrations of a circular membrane (a drum), the study of planetary motion, and the study of wave propagation, diffusion, and potential involving cylindrical geometries.

Derive provides capabilities to evaluate, analyze and plot solutions to these two equations.

Solution: The functions that solve the two Bessel equations are, appropriately enough, called Bessel functions. By using the power series solution technique, a series solution that converges for $n > 0$ and $0 \leq x < \infty$ is found. This solution is denoted by $J_n(x)$ and is called the Bessel function of the first kind of order n. The second linearly independent series solution can be found and defined in a related fashion. It is denoted by $Y_n(x)$ and called the Bessel function of the second kind of order n. Therefore, the general solution of Bessel equation of order n is $y = aJ_n(x) + bY_n(x)$, where a and b are arbitrary constants.

Derive contains functional definitions and approximations for these two functions in the utility file SPECIAL.MTH. This file is loaded into the work area using the **`Transfer, Load`** command. There are numerous special functions in this file. For now, discussion is limited to the functions in Part 5 (Bessel functions) of this utility file. The following table contains the commands in this part of the file and explains their use.

Command	Function
BESSEL_J(n,z)	Uses the integral definition of $J_n(z)$ to determine values of the function; this function is slow and sometimes cannot be evaluated exactly
BESSEL_J_SERIES (n,z,m)	Uses a truncated series to approximate $J_n(z)$; good for $z < 10$ with m increasing as z increases; excellent for $z < 1$ and m as small as 4.
BESSEL_J_ASYMP(n,z)	Uses the asymptotic expression to approximate $J_n(z)$ good for $z > 1$ and excellent for $z > 10$.
BESSEL_Y(n,z)	Uses BESSEL_J to find $Y_n(z)$; therefore is slow and possibly may not be evaluated
BESSEL_Y_ASYMP(n,z)	Uses the asymptotic expression to approximate $Y_n(z)$ good for $z > 1$ and excellent for $z > 10$.
AI(z)	Evaluates the Airy function (a close relative to $J_{1/3}(z)$) which solves Airy's equation: $y'' - xy = 0$
BI(z)	Evaluates the BAiry function, which is the 2nd linearly independent solution of the Airy equation.

Commands in the Bessel function part of utility file SPECIAL.MTH and their functions

First, let's compare values for the three different ways to obtain $J_n(x)$ using Derive commands. If $n = 1$ and $x = 2$, then **Author** the command `BESSEL_J(1,2)` and **Simplify** to obtain

```
61:  BESSEL_J (1, 2)

62:  2 ∫_0^π SIN (t_)^2 COS (2 COS (t_)) dt_
     ----------------------------------------
                       π
```

Unfortunately, Derive is unable to compute this definite integral exactly and just returns the set-up of the integral definition rather than its evaluation and issues a warning beep to alert the user that something is wrong.

This means the symbolic integration for this function is too hard for Derive. Don't laugh or sneer, it's nothing you want to try by hand either and may not be possible at all. The question is, now what? Luckily, there are several alternatives. Probably, the most obvious is to evaluate this integral definition with numerical quadrature. This is done through execution of the **Options, Precision** command. Change the mode to **Approximate** by entering **A** and then **Tab** over and enter the number of digits of accuracy desired. For this example use 5. Now the integral can be approximated by re-execution of the command **Simplify**. Except this time numerical quadrature will be performed. It takes several seconds to obtain the following result.

```
               π
         2 ⌠        2
         2 ⌡  SIN (t_)  COS (2 COS (t_)) dt_
62:        0
       ─────────────────────────────────────
                       π

63:  0.576639
```

Another alternative is to approximate $J_1(2)$ by the truncated series. This evaluation should be quite accurate for $x = 2$. Do this through **Author** and enter `BESSEL_J_SERIES(1,2,m)`. Try $m = 5$ first by putting 5 in for m directly in the command's input. An answer close to that found by the integral approximation is obtained in much less time. The following Derive display shows the command and its result.

```
64:  BESSEL_J_SERIES (1, 2, 5)

65:  0.576724
```

Try $m = 10$ to see if the answer improves. It doesn't, so this value should be an accurate evaluation of $J_1(2)$.

The final alternative is to use the asymptotic formula. Notice that the previous table indicates this technique may not be very accurate in this range where $x = 2$. Enter **Author** and `BESSEL_J_ASYMP(1,2)`. **Simplify** and **approX** to get

```
66:  BESSEL_J_ASYMP (1, 2)

      √2 SIN (2)     √2 COS (2)
67:  ──────────── - ────────────
        2 √π           2 √π

68:  0.528775
```

This answer is a bit different (about 10% relative error) than the previous approximations for $J_1(2)$. This seems to confirm that this asymptotic method may not be appropriate for small values of x (less than about 5).

Let's do the same experiment for the evaluation of $J_0(12.5)$. **Author** `BESSEL_J(0,12.5)` and select **Simplify** to obtain

```
69:  BESSEL_J (0, 12.5)

        π
       ⌠        ┌ 25 COS (t_) ┐
       │  COS   │─────────────│ dt_
70:    ⌡        └      2      ┘
        0
     ──────────────────────────────
                   π

71:  0.146883
```

This calculation takes quite some time. The numerical quadrature has to work hard to get the 5-digit accuracy still required. Next try the truncated series approximations through the command `BESSEL_J_SERIES(0,12.5,m)`. Remember to replace m with the number of terms of the series to be used in the evaluation. Shown here are the commands and results for $m = 10, 20,$ and 100. The calculation for $m = 100$ was performed in **Exact** mode and then approximated.

```
72:  BESSEL_J_SERIES (0, 12.5, 10)

73:  158.412

74:  BESSEL_J_SERIES (0, 12.5, 20)

75:  0.134974

76:  BESSEL_J_SERIES (0, 12.5, 100)

      30995508442065895746849572971813612166540570083801501558371289721247213014
80:  -------------------------------------------------------------------------
      21102023977540044472357546928382900673216413083077750373923516959522145261

81:  0.146884
```

Only the calculation with the summation containing 100 terms is near the answer obtained with the previous integral method, which should be reasonably accurate. This indicates that the truncated series are inaccurate in this region of larger x unless many terms are kept in the series. The asymptotic approximation should do well for this value of x. The display for this operation is as follows.

```
82:  BESSEL_J_ASYMP (0, 12.5)
```

$$83:\quad \frac{\sqrt{2}\,\cos\left[\frac{25}{2}\right]}{5\,\sqrt{\pi}} + \frac{\sqrt{2}\,\sin\left[\frac{25}{2}\right]}{5\,\sqrt{\pi}}$$

```
84:  0.148641
```

The asymptotic solution is close to the other two accurate solutions (integral and 100-term series) and is calculated in much less time. This experiment supports the tables suggestion that the asymptotic approximation is the most efficient technique for larger values of x.

Let's try plotting a Bessel function. The function we will use for this experiment is $J_{1/2}(x)$. Remember that the function evaluations for individual points for the three techniques (especially quadrature and truncated series) take quite some time. Therefore, a plotting window is opened so the plots are not redrawn, and even then the plotting will be slow.

First, open a plot window using the **Window, Split** command. Select **Horizontal** and split at line #10. Move to the larger window by **Window, Next** and execute **Window, Designate, 2-D plot**. Now the screen will keep the plot window open and the three functions can be plotted in sequence. Move back and forth from the algebra window to the plot window using **Window, Next** commands or the **Algebra** and **Plot** commands.

Next, go back to the algebra window, **Author** the three functions (one at a time), (i) `BESSEL_J(1/2,x)`, (ii) `BESSEL_J_SERIES(1/2,x,3)`, and (iii) `BESSEL_J_ASYMP(1/2,x)`. Highlight the first function and move back to the plot window. There is no need to plot values for $x < 0$; therefore, before plotting in the default window which centers the plot on the origin, use the direction keys or the **Move** command to move the cross to the right a couple of units, then execute the **Center** command. This should change the horizontal plotting interval to show more of $x > 0$. Some finer tuning can be used to get just the positive x-axis to show in the window. When the interval is acceptable, use the **Plot** command to get the plot of the first highlighted function (*Warning: some of these plots take considerable time to produce*). When the first plot is done, move back to the algebra window and highlight the next function. Then go back to the plot window and **Plot** that function. The display, after all three functions are plotted, looks like this.

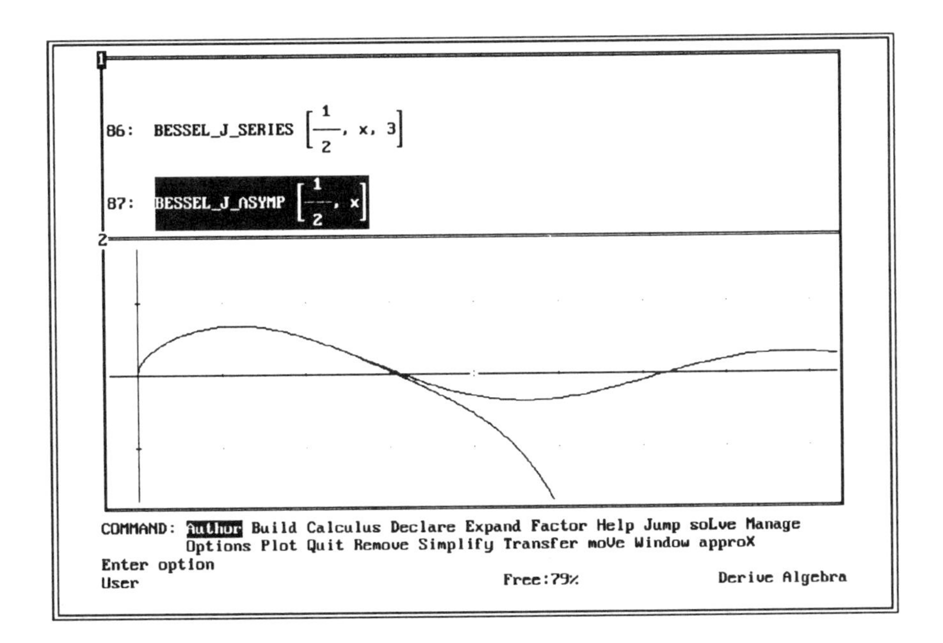

The 3-term series approximation doesn't do very well for larger x.

Similar evaluations, plots and experiments can be done with the Bessel functions of the second kind $Y_n(x)$ and the Airy functions $Ai(x)$ and $Bi(x)$.

However, we won't do them here.

Now, let's use this information solve the differential equation

$$x^2y'' + xy' + (x^2 - 0.01)y = 0$$

with boundary conditions $y(1) = 3$ and $y(10) = 6$.

This is a Bessel equation of order 0.1 and the general solution is $y = aJ_{0.1}(x) + bY_{0.1}(x)$. The boundary conditions require the two equations

$$3 = aJ_{0.1}(1) + bY_{0.1}(1)$$

$$6 = aJ_{0.1}(10) + bY_{0.1}(10)$$

to be satisfied. The four Bessel functions are evaluated using four appropriate functions in the previous table as shown in this Derive display.

```
88:  BESSEL_J_SERIES (0.1, 1, 25)

89:  0.770765

92:  BESSEL_Y_ASYMP (0.1, 1)

93:  0.0458707

94:  BESSEL_J_ASYMP (0.1, 10)

95:  -0.235487

96:  BESSEL_Y_ASYMP (0.1, 10)

97:  0.0905946
```

The two initial conditions are easily converted to a system of equations as components of a vector using the **Author** command to get

```
99:  [3 = a 0.770765 + b 0.0458707, 6 = a (-0.235487) + b 0.0905946]
```

The equations are solved for a and b using the commands **soLve** and **approX**. These commands produce the following result.

```
             34404000000          53310510000000
100: [a = - ───────────, b = ──────────────]
             806291003999          806291003999

101: [a = -0.0426694, b = 66.1182]
```

So the solution to the equation is $y = -0.04266J_{0.1}(x) + 66.1182Y_{0.1}(x)$. This function can be *Authored* into the work area with the expression `-0.04266 BESSEL_J(0.1,x) + 66.1182 BESSEL_Y(0.1,x)`. Then move into the plot window, clear all previous plots, and adjust the plot interval to show $5 < x < 9$ accurately and set the y-scale to 10. Then **Plot** the expression, whose graph appears as follows:

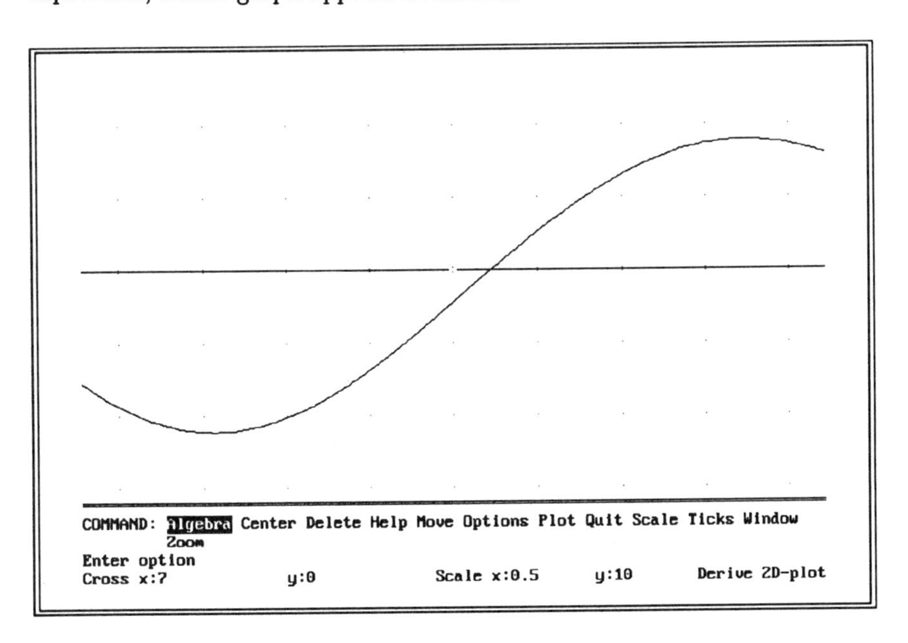

Example 2.15: Difference Equation

I must study politics and war that my sons may have liberty to study mathematics and philosophy.

—John Adams [1780]

Subject: Using Difference Equations for a Model in Personal Finance

Problem: This problem is an exploration into several different scenarios involving difference equation models for finance. Difference equations are the discrete mathematics analog to differential equations. This problem involves several scenarios where first-order difference equations are used as models for accumulating and withdrawing money from savings accounts.

If $D(n)$ is the number of dollars in a savings account after n compounding periods, the account collects interest at a $100i$ percent annual rate compounded m times a year, and a dollars are added to the account at the end of each compounding period, then the difference equation model is

$$D(n+1) = (1 + i/m)D(n) + a.$$

If a is negative, then dollars are withdrawn from the account.

Solution: Derive does not perform automatic iteration of discrete functions, However, the Derive utility file RECUREQN.MTH has several command functions that solve some of the common difference equations. Other names commonly used for these types of equations are discrete dynamical systems and recurrence relations. The following table provides a short description of some of the commands in this file.

Command	Function
LIN1_DIFFERENCE (p,q,n,n0,y0)	Solves the first-order nonhomogeneous linear difference equation $y(n+1) = p(n)y(n) + q(n)$ with $y(n0) = y0$.
GEOMETRIC1 (p,q,k,n,n0,y0)	Solves the linear difference equation $y(kn) = p(n)y(n) + q(n)$, $y(n0) = y0$.
LIN2_RED_CCF_RECUR_DISC (p,q,n)	Evaluates the discriminant function for the second-order, constant coefficient difference equation $y(n+2) + py(n+1) + qy(n) = 0$.
LIN2_RED_CCF_RECUR_POS (p,q,n)	Solves the 2nd-order difference equation when the discriminant is positive.
LIN2_RED_CCF_RECUR_NEG (p,q,n)	Solves the 2nd-order difference equation when the discriminant is negative.
LIN2_RED_CCF_RECUR_0 (p,q,n)	Solves the 2nd-order difference equation when the discriminant is 0.
LIN2_COMPLETE(u,v,r,n)	Solves for the particular solution of $y(n+2) + py(n+1) + qy(n) = r(n)$ where u and v are complementary solutions.
IMPOSE_BV2 (n,y,n1,y1,n2,y2)	Solves for the arbitrary constants @1 and @2 in the solution of a 2nd-order difference equation with conditions $y(n1) = y1$ and $y(n2) = y2$.

Commands for solving difference equations in utility file RECUREQN.MTH and their functions

Therefore, the command to solve the above first-order equation with condition $D(k) = D_k$ is **LIN1_DIFFERENCE(**$1 + i/m, a, n, k, D_k$**)**.

Now, let's tackle the first scenario. If there are 2 banks offering competing savings plans where one plan had a 5.125% interest compounded yearly and the other plan had a 5% interest compounded daily, which bank has a better deal for the customer?

The model for the daily compounding in this scenario is

$$D(n+1) = (1 + 0.05/365)D(n) \quad \text{with } D(0) = c.$$

Once the utility file is loaded into the work area with commands **Transfer**, **Load** and filename **RECUREQN**. This difference equation is entered into the work area with **Author** and **LIN1_DIFFERENCE(1+0.05/365,0,n,0,c)**.

Simplify to get its solution.

```
                                0.05
34:  LIN1_DIFFERENCE  [1 + ———————, 0, n, 0, c]
                                365

         ┌ 7301 ┐n
35:  c   │——————│
         └ 7300 ┘
```

We want to know the balance after 1 year or 365 days, so **Manage, Substitute** for n the value 365. Just hit **Enter** when the value for the variable c is asked for. It is probably best to change the arithmetic precision to **Approximate** through the **Options, Precision** commands before the next operation. This will save the computation of a fraction with hundreds of digits in the numerator and denominator. Then, **Simplify** this expression to obtain

```
         ┌ 7301 ┐365
36:  c   │——————│
         └ 7300 ┘

37:  1.05126 c
```

From this result, we see that the effective interest rate is higher for the bank with the daily compounding, but lower numerical rate.

The next scenario involves developing a lifetime savings/retirement plan. You want to retire in 35 years and in that time save enough so you can withdraw the equivalent of \$30,000 in current money for the next 25 years of retirement. If 6% annual inflation is assumed, the difference equation model to determine the equivalent of \$30,000 in 35 years is

$$D(n+1) = (1 + 0.06/1)D(n) \quad \text{with } D(0) = 30,000.$$

To solve this, **Author** and **Simplify** the command

```
LIN1_DIFFERENCE(1+0.06,0,n,0,30000)
```
.

Then **Manage, Substitute** 35 for n and **Simplify**. Derive responds with the following output.

```
38:  LIN1_DIFFERENCE (1 + 0.06, 0, n, 0, 30000)

            0.0582689 n
39:  30000 ê

            0.0582689 35
40:  30000 ê

                  5
41:  2.30582 10
```

Wow! You'll need over \$230,000 a year for each of your 25 years (or a total of \$5,764,500) to retire. Can you save enough to have that much in 35 years? How much do you need to save?

Assume you will invest all your savings when you retire at 10.2% interest compounded annually and withdraw \$230,00 per year for 25 years from this account. At that time you will have exhausted your savings and the balance will be \$0. The difference equation model for this plan is

$$D(n+1) = (1 + 0.1/1)D(n) - 230000 \quad \text{with} \quad D(25) = 0.$$

Once again, **Author** and **Simplify** the expression

LIN1_DIFFERENCE(1+0.1,-230000,n,25,0).

The resulting Derive display is shown.

```
42:  LIN1_DIFFERENCE (1 + 0.1, -230000, n, 25, 0)

            6              5  0.0953101 n
43:  2.3 10   - 2.1228 10  ê
```

What is $D(0)$? **Manage, Substitute** 0 for n and **Simplify** to get

```
            6              5  0.0953101 0
44:  2.3 10   - 2.1228 10  ê

                  6
45:  2.08772 10
```

Surprise! You need to save \$2,000,000 during the next 35 years. At least this is considerably better than \$5,764,500. How much do you have to save each month? Assume you will invest in a savings account that pays 8% interest compounded monthly and will deposit p dollars every month for 420 months (35 years). A practical model for this is

$$D(n+1) = (1 + 0.08/12)D(n) + b \quad \text{with} \quad D(420) = 2,000,000.$$

Author this into Derive with the command

```
LIN1_DIFFERENCE(1 + 0.08/12,b,n,420,2000000)
```

Simplify this expression to get the solution. **Manage, Substitute** 0 for n. Return to the **Exact** mode using **Options, Precision** and **soLve** for b. Then, **approX** this result. The expressions in the work area to do these steps are shown.

```
46:  LIN1_DIFFERENCE [1 + 0.08/12, b, n, 420, 2000000]

                                  0.00664451 n
47:  3.06891 (3 b + 40000) ê                   - 150 b

                                  0.00664451 0
48:  3.06891 (3 b + 40000) ê                   - 150 b

           476020000
49:  b =  -----------
            545961

50:  b = 871.893
```

Now you know that you need to save about \$870 per month for the next 35 years to retire on the equivalent of \$30,000 per year for the following 25 years.

Example 2.16: Budget Growth

> ***But the velocities of the velocities—the second, third, fourth, and fifth velocities, etc.—exceed, if I mistake not, all human understanding.***
>
> —George Berkeley [1734]

Subject: Budget Growth and Conversion of a System of Equations to a Higher-Order Equation

Problem: An organization's budget is divided into two categories, procurement and operations. Given the type of organization and by letting $e_1(t)$ be the procurement expenditure function, $e_2(t)$ be the operations expenditure function, and $g(t)$ be the independent growth factor, the time rate of change of the expenditures is modeled by the following system of differential equations:

$$\frac{de_1}{dt} = 0.1e_1 - 0.01e_2 + g(t)$$

and

$$\frac{de_2}{dt} = 0.3e_1 + 0.02e_2 + g(t).$$

If the current expenditures are $e_1(0) = 250$ and $e_2(0) = 114$ and if the growth function is $g(t) = 2$, solve the differential model and analyze the future patterns in the organization's expenditures.

Solution: One way to solve a system of 2 differential equations is to convert it into a second-order equation with only one dependent variable by differentiating one of the equations and substituting the other into it. If this is done for this problem, we can use Derive's commands in ODE2.MTH to help solve the resulting second-order equation. Probably the easiest and best way to perform the manipulations to make the conversion is to use Derive.

Derive does not allow subscripted variable or function names, so we can't use $e_1(t)$ or $e_2(t)$ directly in Derive. However, in **`Word`** input mode, Derive does allow two or more characters in a name like $E1$ and $E2$, so we'll use these names in place of the subscripted names used above. This mode is established with the **`Options, Input, Word`** menu command.

Start the conversion work by using the **`Declare, Function`** command to make first $E1$ and then $E2$ functions of t. This command queries the user for the function name and its variables. The keystrokes to do this are `d f e1 Enter Enter t Enter`

and

`d f e2 Enter Enter t Enter`.

After these declarations, the screen display will show

```
1:    E1 (t) :=

2:    E2 (t) :=
```

Next **Author** the two equations. The inputs to do this are

`DIF(e1(t),t) = 0.1e1(t) - 0.01e2(t) + 2`

and

`DIF(e2(t),t) = 0.3e1(t) + 0.02e2(t) + 2`.

The following image of the work area shows these operations.

```
        d
3:      — E1 (t) = 0.1 E1 (t) - 0.01 E2 (t) + 2
        dt

        d
4:      — E2 (t) = 0.3 E1 (t) + 0.02 E2 (t) + 2
        dt
```

The simplest Derive command to differentiate the entire equation for $\frac{de_1}{dt}$ is `DIF(#n,t)` where n is the number of the equation in the work area (#3 in the work area for the screen above). Use **Simplify** to get the result shown.

$$5: \frac{d}{dt}\left[\frac{d}{dt} E1(t) = 0.1\ E1(t) - 0.01\ E2(t) + 2\right]$$

$$6: \left[\frac{d}{dt}\right]^2 E1(t) = \frac{\frac{d}{dt} E1(t)}{10} - \frac{\frac{d}{dt} E2(t)}{100}$$

Now, substitute the right-hand side of the equation for $\frac{de_2}{dt}$ into the new second-order equation. Unfortunately, Derive does not allow substitution for a declared function (or its derivatives) through use of the **Manage, Substitute** command. Therefore, the substitution must be done by hand. The keystroking hints of Section 1.3 are very helpful. Execute the **Author** command and use the direction keys to move the highlight, the **F3** or **F4** key to copy highlighted text from the work area to the command line, and the **Ctrl-s** and **Ctrl-d** keystrokes to move left and right in the command line. If you have not done this kind of keystroking, now is the time to start. The alternative is retyping the entire input line. The result of this substitution is as follows.

$$8: \left[\frac{d}{dt}\right]^2 E1(t) = \frac{\frac{d}{dt} E1(t)}{10} - \frac{0.3\ E1(t) + 0.02\ E2(t) + 2}{100}$$

This still leaves $e_2(t)$ in the equation, so this substitution procedure must be repeated for $e_2(t)$ in terms of $e_1(t)$ and $\frac{de_1}{dt}$ from the first equation, giving the following result. The line is truncated on the right by the screen width used by Derive.

$$10: \left[\frac{d}{dt}\right]^2 E1(t) = \frac{\frac{d}{dt} E1(t)}{10} - \frac{0.3\ E1(t) + 0.02\left[- 100\ \frac{d}{dt} E1(t) + 10\ E1(t\right.}{100}$$

Simplify this equation to obtain the second-order equation with constant coefficients for $e_1(t)$ as shown.

```
                                   d
                              24 ── E1 (t) - E1 (t) - 12
11:   ┌d ┐2                       dt
      │──│  E1 (t) = ──────────────────────────────────
      └dt┘                            200
```

Now the commands in utility file ODE2.MTH can be used to help solve this equation. Load this utility file into the work area with the **Transfer, Merge** command. This leaves the previous commands in the work area in case any of them are needed again. These utility file commands could be placed in another window and then used in the current window. The equation has constant coefficients and can be expressed in the form used by Derive's file ODE2 as $y'' + py' + qy = r(x)$ with $E1$ substituted for y, $p = -24/200$, $q = 1/200$, and $r = -12/200$. Also the independent variable for this problem is t instead of x. Therefore, the in-line command from ODE2 to determine the value of the discriminant function is `LIN2_RED_CCF_DISC(- 24/200,1/200)`. The display of this command with the subsequent execution of the **Simplify** command gives

```
                           ┌   24    1 ┐
46:  LIN2_RED_CCF_DISC     │- ───, ─── │
                           └   200  200┘

         7
47:  - ────
       1250
```

This negative (< 0) result determines the next command needed to obtain the complementary solution. **Author** and **Enter** the command `LIN2_RED_CCF_NEG(- 24/200,1/200,t)` and **Simplify**. The result is

```
48:  LIN2_RED_CCF_NEG [- 24/200, 1/200, t]

49:  [@2 COS [√14 t/100] + @1 SIN [√14 t/100]] ê^(3 t / 50)
```

The particular solution is determined by the **LIN2_COMPLETE** command with its arguments as shown here in the display of its command and result.

```
51:  LIN2_COMPLETE [COS [√14 t/100] ê^(3 t / 50), SIN [√14 t/100] ê^(3 t / 50), - 12/200, t

52:  -12
```

This particular solution is a simple constant since the forcing term $r(t)$ is a constant. Now the solution for e_1 can be assembled from the complementary and particular solutions using keystroking as

```
53:  [@2 COS [√14 t/100] + @1 SIN [√14 t/100]] ê^(3 t / 50) - 12
```

The function for e_2 can be found from substitution of e_1 and $\frac{de_1}{dt}$ into the first equation of the system. Using Derive to do this, **Author** the command
`DIF(#53,t) = 0.1(#53) - 0.01e + 2`. The variable **e** is used in place of the function **E2(t)**. **Simplify** and **soLve** to get the following display as shown with truncated expressions.

$$54: \frac{d}{dt}\left[\left[@2\ \mathrm{COS}\left[\frac{\sqrt{14}\ t}{100}\right] + @1\ \mathrm{SIN}\left[\frac{\sqrt{14}\ t}{100}\right]\right] \hat{e}^{3\ t\ /\ 50} - 12\right] = 0.1\left[\left[@2\ \mathrm{COS}\left[-\right.\right.\right.$$

$$55: \left[\left[\frac{\sqrt{14}\ @1}{100} + \frac{3\ @2}{50}\right] \mathrm{COS}\left[\frac{\sqrt{14}\ t}{100}\right] + \left[\frac{3\ @1}{50} - \frac{\sqrt{14}\ @2}{100}\right] \mathrm{SIN}\left[\frac{\sqrt{14}\ t}{100}\right]\right] \hat{e}^{3}$$

The solution for e_2 after issuing **Expand** and **approX** is

$$60: e = -\frac{22536}{6023}\ @1\ \mathrm{COS}\left[\frac{1822}{48695}\ t\right] \hat{e}^{0.06\ t} + 4\ @2\ \mathrm{COS}\left[\frac{1822}{48695}\ t\right] \hat{e}^{0.06\ t}$$

$$+ 4\ @1\ \mathrm{SIN}\ (0.0374\ t)\ \hat{e}^{0.06\ t} + 3.74\ @2\ \mathrm{SIN}\ (0.0374\ t)\ \hat{e}^{0.06\ t} + 80$$

To find @1 and @2 that satisfy the initial conditions, just use the functional definitions of `E1(t)` and `E2(t)` and place the 2 equations in a vector. **Simplify** and **soLve**. The display of the steps to do this is as follows:

$$61: E2\ (t) := \left[\sqrt{2}\ (2\ \sqrt{2}\ @1 + \sqrt{7}\ @2)\ \mathrm{SIN}\left[\frac{\sqrt{14}\ t}{100}\right] - \sqrt{2}\ (\sqrt{7}\ @1 - 2\ \sqrt{2}\ @2)\ \mathrm{COS}\left[-\right.\right.$$

$$62: E1\ (t) := \left[@2\ \mathrm{COS}\left[\frac{\sqrt{14}\ t}{100}\right] + @1\ \mathrm{SIN}\left[\frac{\sqrt{14}\ t}{100}\right]\right] \hat{e}^{3\ t\ /\ 50} - 12$$

$$63: [E1\ (0) = 250,\ E2\ (0) = 114]$$

$$64: [@2 - 12 = 250,\ -\sqrt{14}\ @1 + 4\ @2 + 80 = 114]$$

$$65: \left[@1 = \frac{507\ \sqrt{14}}{7},\ @2 = 262\right]$$

Once the solutions are verified, they can be plotted. Highlight the right-hand side of the function definitions for e_1 and e_2. Then the **Plot** menu command is selected to open the plotting window. Before the plotting is started, use **Scale** to set x-scale to 1.5 and y- scale to 1000. Then move the movable cross with the right direction arrow (→) or the **Move** command. The **Center** command is executed to move the plot range to positive t values. Now, the **Plot** command is issued for both components of the solution to obtain the following graph.

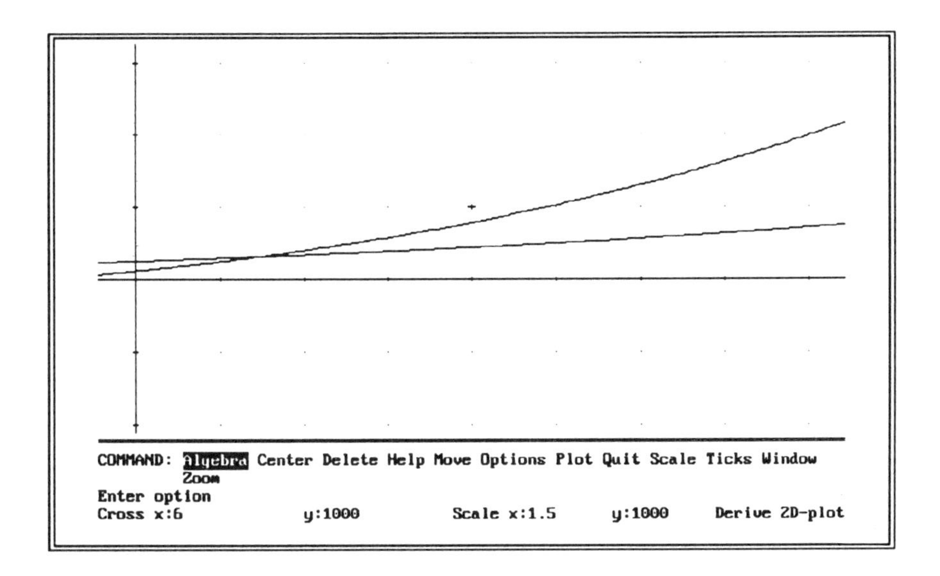

Therefore, even though procurement is currently the big spender, the organization would do better to think about long-range cost controls for its operating expenditures.

Example 2.17: Taylor Polynomial

...over the entrances to the gates of the temple of science are written the words: Ye must have faith.

—Max Planck [1932]

Subject: Taylor Polynomial Approximations

Problem: Approximate the solution to the nonlinear differential equation

$$y' = \sin(xy) \quad \text{with} \quad y(a) = b$$

and analyze the results.

Solution: An approach to approximating solutions to nonlinear differential equations is to expand the functions into Taylor polynomials accurate in some neighborhood of the point of expansion. Derive has a 4th-degree truncated Taylor-series solution operator in the utility file ODE1.MTH to perform this kind of approximation. Once the file ODE1 is loaded into the work area with the **Transfer, Load** command, the command to approximate the solution to this problem is

```
TAY_ODE1(sin(xy),x,y,a,b)
```

Simplify this operation to obtain the following result which shows the output wrapped around and some truncation of terms. Remember Derive keeps all its output for an expression on one line. In order to access the part of the expression off the screen, the direction arrows must be used.

68: TAY_ODE1 (SIN (x y), x, y, a, b)

$$69:\quad \left[\frac{a^3 (x-a)^4 \operatorname{SIN}(a\ b)}{12} + \frac{a^2\ b\ (x-a)^4}{12}\right] \operatorname{COS}(a\ b)^3 +$$

$$+ \left[\frac{a\ (x-a)^3\ (11\ x - 3\ a) \operatorname{SIN}(a\ b)}{24} + \frac{b\ (x-a)^3\ (7\ x + a)}{24}\right] \operatorname{COS}(a\ b)^2$$

$$\left[\frac{a^3\ (x-a)^4 \operatorname{SIN}(a\ b)^3}{6} + \frac{5\ a^2\ b\ (x-a)^4 \operatorname{SIN}(a\ b)^2}{12} +\right.$$

$$\frac{(x-a)^2\ (a\ x^2\ (a^2 + 7\ b^2) - 2\ x\ (a^4 + 7\ a^2\ b^2 + 4) + a\ (a^4 + 7\ a^2\ b^2 - 4}{24}$$

$$\left.+ \frac{b\ (x-a)^2\ (x^2\ (a^2 + b^2) - 2\ a\ x\ (a^2 + b^2) + a^4 + a^2\ b^2 - 12)}{24}\right]$$

This approximation can be plotted for various values of parameters a and b. The set of initial conditions $y(0) = -3, -2, -1, 0, 1, 2, 3$ can be entered one at a time using the **Manage, Substitute** command. Then **Simplify** each expression to obtain these seven approximations for the seven respective initial conditions.

$$78:\quad \frac{3(x^4 - 2x^2 - 4)}{4}$$

$$79:\quad \frac{x^4 - 12x^2 - 24}{12}$$

$$80:\quad -\frac{x^4 + 6x^2 + 12}{12}$$

$$81:\quad 0$$

$$82:\quad \frac{x^4 + 6x^2 + 12}{12}$$

$$83:\quad -\frac{x^4 - 12x^2 - 24}{12}$$

$$84:\quad -\frac{3(x^4 - 2x^2 - 4)}{4}$$

The easiest way to plot these functions all at once is to place them in a vector as components through **Author** of the input

```
[#78, #79, #80, #81, #82, #83, #84]
```

.

Then go to the **Plot** window, **Scale** the y-scale to 1.5, and execute **Plot**. The plots of these functions reveal symmetry with respect to both axes and provide visualization of the solution behavior near the initial point $x = 0$.

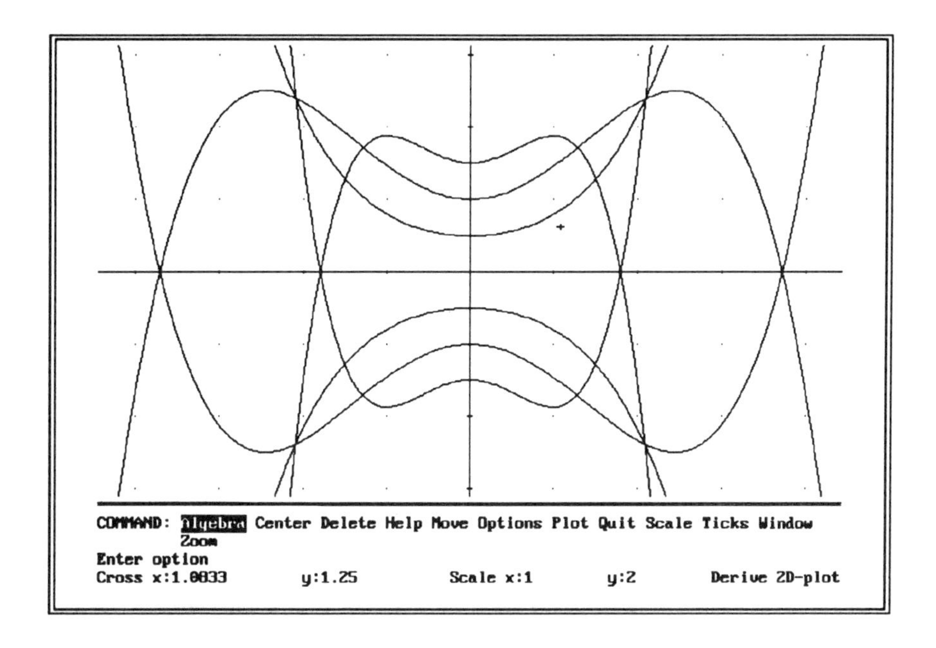

However, the solution trajectories cross one another away from their start points, which shows the poor nature of the approximation away from $x = 0$. This technique is only good for analyzing local behavior. Other values of a and b could be entered into the approximation function to check local behavior in other regions.

Example 2.18: Picard Iteration

A collection of facts is no more a science than a heap of stones is a house.

—Jules Henri Poincare [1903]

Subject: Successive Iteration (Picard's Method)

Problem: Find an approximation to the solution of

$$y' = x^2 - 2y^2 - 1 \text{ with } y(0) = 0.$$

Solution: This is a nonlinear differential equation which is not separable or exact and therefore does not lend itself easily to a simple closed-form solution. The approximation method of successive iteration can be used to approximate the solution for such problems. The method needs an initial approximation and the initial condition to get started. In this case, let's start with $y = x$ as the initial approximation to the solution. This approximation is the linear polynomial that satisfies the initial condition. Many other initial approximations are possible for this problem.

Derive's successive iteration operator is in the utility file ODE1.MTH. Load this file into the work area first with the **Transfer, Load** command. The command PICARD($r(x, y), y_{prev}, x, y, x_0, y_0$) produces the next iterative approximation for the differential equation $y' = r(x, y)$ with approximation y_{prev} and initial condition $y(x_0) = y_0$. Therefore, for this problem the first iteration is performed by

```
PICARD(x^2-2y^2-1,x,x,y,0,0)
```

Simplify this operation to obtain the next approximation to the solution as shown

```
                  2       2
68:  PICARD (x  - 2 y  - 1, x, x, y, 0, 0)

           3
          x
69:  -  ———— - x
           3
```

In entering the command for the next iteration, take advantage of the previous expressions and use the highlight and **F3** key to produce an input line and working display as shown.

```
AUTHOR expression: PICARD (x^2 - 2 y^2 - 1, - x^3 / 3 - x, x, y, 0, 0)

Enter expression
Simp(68)              D:2.17              Free:54%              Derive Algebra
```

$$70:\quad \text{PICARD}\left[x^2 - 2y^2 - 1,\ -\frac{x^3}{3} - x,\ x,\ y,\ 0,\ 0\right]$$

Once again, the **Simplify** command produces the next successive approximation

$$71:\quad -\frac{x\,(10x^6 + 84x^4 + 105x^2 + 315)}{315}$$

Two more iterations produce the following display with some truncation.

$$72:\quad \text{PICARD}\left[x^2 - 2y^2 - 1,\ -\frac{x\,(10x^6 + 84x^4 + 105x^2 + 315)}{315},\ x,\ y,\ 0,\ 0\right]$$

$$73:\quad -\frac{x\,(5720x^{14} + 110880x^{12} + 714168x^{10} + 2282280x^8 + 7837830x^6 + 113513}{42567525}$$

$$74:\quad \text{PICARD}\left[x^2 - 2y^2 - 1,\ -\frac{x\,(5720x^{14} + 110880x^{12} + 714168x^{10} + 2282280x}{42}\right.$$

$$75:\quad -\frac{x\,(2114665444432000x^{30} + 8763801830784000x^{28} + 151861523283500800x^{26} +}{}$$

Plots of the second approximation and the last approximation to the solution are made using the **Plot** window. **Scale** the y-scale to 5 and execute **Plot** to obtain the following.

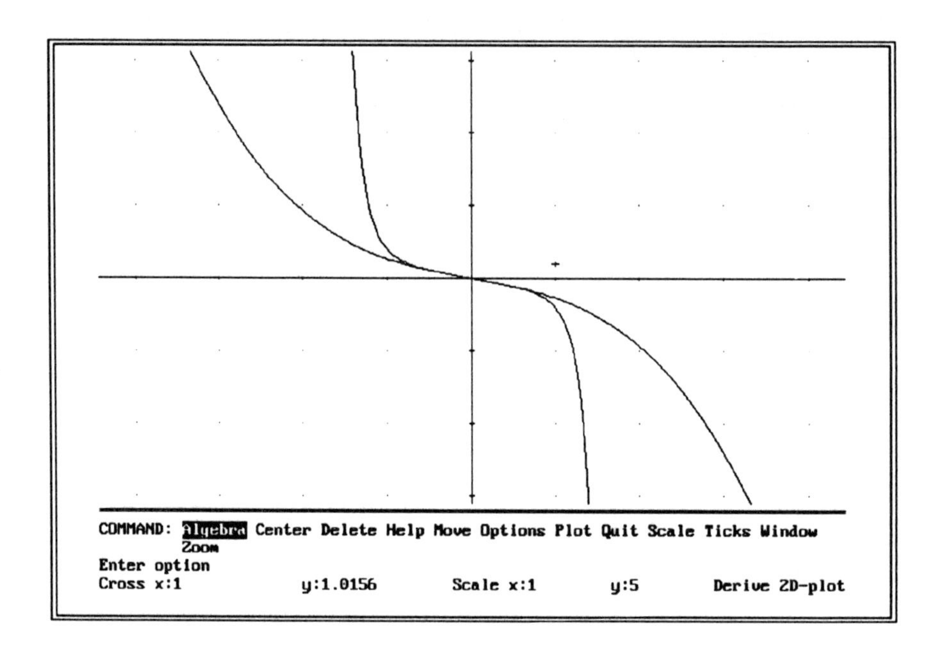

Chapter 3

Exercises

We shall not cease from exploration
And the end of all our exploring
Will be to arrive where we started
And know the place for the first time.

—T. S. Eliot *Little Gidding* [1942]

These exercises are to be done by the reader. These problems direct you to learn new things about differential equations, Derive, and mathematics. Some of the problems, solution techniques, and Derive commands are similar to those in the examples of Chapter 2 and refer to those examples as a source of help. But most of the exercises in this chapter lead you to explore the mathematics and software and discover new results on your own. As you solve the questions asked in the exercises,ask yourself more "what if" questions and solve them. Do not give up until you have satisfied yourself. Good luck and have fun exploring.

Exercise 3.1

Science can tell us how to do many things, but it cannot tell us what ought to be done.

— Anonymous

Subject: A Model for Waste Disposal in the Ocean

Purpose: To analyze a plan for dumping toxic waste in the ocean using a first-order linear differential equation.

Given: One method proposed for disposing toxic waste is to place it in sealed drums and dump the drums in a deep part of the ocean. Tests showed that drums could be made that would never leak from corrosion, but there still was concern over their breaking open from impact with the ocean floor. Through the summation of the forces of gravity, buoyancy, and drag, a model for the velocity ($v(t)$) of the drum's descent in time through the water is developed as

$$\frac{dv}{dt} + \frac{ag}{wz}v = \frac{g}{w}(w - b),$$

where a is the coefficient of drag for the drums, g is the acceleration of gravity (32.2 ft/sec^3), w is the density of the toxic waste, b is the density of the water (produces buoyancy), and z is the volume of the drum.

Exercises:

1. Solve the differential equation model using 2 different commands in file ODE1.MTH. What is the maximum velocity of the drums?

2. If 55-gallon drums (7.4 ft^3) are dumped from rest by rolling them off the side of a barge, the waste has density 80 lbs/ft^3, water has density 62.5 lbs/ft^3, and the drag coefficient for this style drum is 0.10, plot the velocity for $0 < t < 30$. What is the maximum velocity of the drums for these values of the parameters?

3. Integrate the function in #1 for velocity to obtain an expression for the distance traveled (y) by the drums in terms of time t.

4. If the parameter values are as given in #2 and if the ocean depth is 400 feet, how fast are the drums traveling at impact.

5. Does it reduce the speed at impact if smaller drums are used? (Assume the drag coefficient is unchanged).

6. If the maximum safe impact velocity is 75 ft/sec, what is the maximum ocean depth for safe dumping?

7. If the density of the waste is increased to 90 lbs/ft^3, what is the maximum ocean depth for safe dumping?

Exercise 3.2

Population, when unchecked, increases in a geometrical ratio. Subsistence increases only in an arithmetical ratio. A slight acquaintance with numbers will show the immensity of the first power in comparison of the second.

—Thomas Malthus [1798]

Subject: Population Growth

Purpose: Analyze three different models for population growth of a specie in a limited-resource environment.

Given: There are several classical first-order differential equation models for population growth. Three are provided. The first model is the logistics equation

$$\frac{dP_1}{dt} = r_1(M - P_1)P_1 \quad \text{with} \quad r, M, P_1 \geq 0.$$

The second model is similar to the first one; its equation is

$$\frac{dP_2}{dt} = -r_2(N - P_2)P_2 \quad \text{with} \quad r, N, P_2 \geq 0.$$

Finally, the third model is

$$\frac{dP_3}{dt} = -r_3(M - P_3)(N - P_3)P_3 \quad \text{with} \quad r, M, N, P_3 \geq 0.$$

Exercises:

1. Plot the phase plane curves (dP_i/dt verses P_i, $i = 1, 2, 3$) for the 3 models with parameter values $r_1 = r_2 = 0.5$, $r_3 = 0.005$, $M = 500$, and $N = 800$. (Hints: use y for dP_i/dt and x for P_i; use x-scale ≈ 200 and y-scale $\approx 100,000$).

2. Use the cross on the plot screen to approximate the equilibrium points for each of the 3 models and then classify these points as stable or unstable.

3. If $P_i(0) = 600$, for $i = 1, 2, 3$, solve the three equations above using the operation SEPARABLE in the utility file ODE1.MTH. See Section 1.12 and Example 2.1 for use of this operation in solving separable equations.

4. Begin a new plot screen either by deleting the stability curves or by overlaying a new screen. Plot the solutions obtained in #3 on the same axes (P_i verses t). Which model predicts the most population when $t = 0.1$. Could this result have been predicted from the stability curves?

Exercise 3.3

I do not love ... to be dunned and teezed by forreigners about mathematical things ...

—Issac Newton [1699]

Subject: Newton's Law of Cooling

Purpose: Solve and analyze models for cooling of a space capsule after splash down.

Given: The model for the cooling of the exterior of a space capsule is

$$\frac{dT}{dt} = -k(T - \beta), t > 0, T(0) = \alpha,$$

where T is the temperature of the capsule, t is the time after splash down, β is the temperature of the ocean, α is the initial temperature of the capsule, and k is a constant of proportionality.

Exercises:

1. Solve for the temperature T as a function of t and the parameters α, β, and k. (Hint: consider using an operation from utility file ODE1.MTH).

2. If the capsule temperature was 650° F at splash down, the ocean is 40° F, and after 20 seconds the capsule temperature has dropped to 450° F, find the value of k.

3. Plot the solution (T verses t) in the above scenario. The domain of interest is $0 \leq t \leq 200$ seconds.

4. What changes are needed in the model if the capsule is lifted out of the water at $t = 100$ seconds and remains at an air temperature of 80° F?

5. Solve for T in the domain of interest $100 \leq t \leq 200$ in the above scenario (#4). Plot this new solution on the same axes as the previous solution. (Hint: it may be helpful to use the **STEP** function to plot the new function for $t > 100$). What is the temperature difference of the capsule at $t = 200$ seconds between the 2 different scenarios?

Exercise 3.4

> *The science of physics does not only give us an opportunity to solve problems, but it helps us also to discover the means of solving them, ...*
>
> —Henri Poincare

Subject: Electrical Circuit—Modeled as a Nonhomogeneous System and Solved Using Variation of Parameters

Purpose: Model an electrical circuit with a system of equations and solve the system using variation of parameters (See Example 2.6)

Given: The following electrical circuit diagram

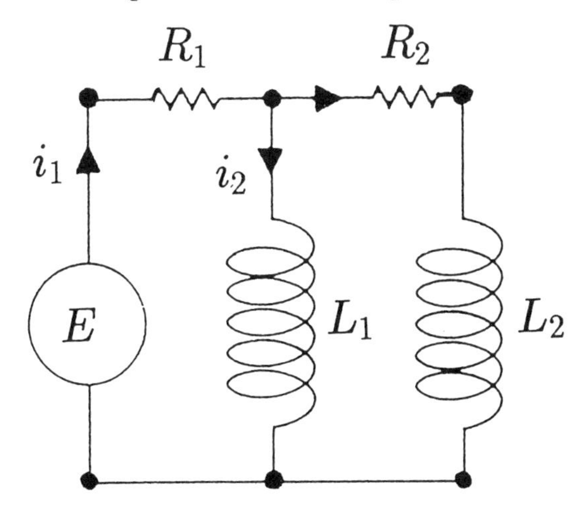

Exercises:

1. Use Kirchoff's first and second laws to write a first-order differential system in matrix notation for i_1 and i_2 in terms of R_1, R_2, L_1, L_2, and E, with $\vec{I} = \begin{bmatrix} i_1 \\ i_2 \end{bmatrix}$.

2. If $R_2 = 3$ ohms, $L_1 = 1$ henry, and $L_2 = 1$ henry, find the eigenvalues and eigenvectors for this system in terms of R_1.

3. Write the complementary solution $\vec{I}_c$.

4. Is there a value of R_1 which produces a positive eigenvalue? Why or why not?

5. If $E = 50$ volts, find the particular solution $\vec{I}_p$ using variation of parameters. (See Example 2.8).

6. Write the general solution $\vec{I}_g$.

7. If $i_1(0) = 0$ and $i_2(0) = 0$, write the solution $\vec{I}$.

8. What happens if $R_1 = 0.1$? if $R_1 = 4$ ohms? if $R_1 = 8$ ohms? Graph i_1 for these three cases on one set of axes $(0 \leq t \leq 10)$.

Exercise 3.5

It is a great nuisance that knowledge can only be acquired by hard work.

— W. Somerset Maugham

Subject: Solving a Complete Second Order Differential Equation with Constant Coefficients using Undetermined Coefficients

Purpose: Find the solution of an initial value problem with parameters using the technique of undetermined coefficients and analyze the effect of varying the parameters.

Given:

$$L[y] = y'' - 2y' + 10y = 3xe^{2x}\cos(3x), \tag{3.1}$$

with initial conditions $y(\pi/2) = 0$ and $y'(\pi/2) = a$.

Exercises:

1. Formulate the auxiliary equation and solve for its roots using Derive. Write the complementary solution, y_c, from these roots.

2. Using the method of undetermined coefficients, write the form of a particular solution, y_p.

3. Using Derive, define the function operator $L[y]$, evaluate $L[y_p]$, and equate it to the right-hand side of the given nonhomogeneous equation (3).

4. Using Derive, solve for the undetermined coefficients by equating the like coefficients.

5. Write the general solution, y_g, from the functions y_c and y_p. Check your solution using Derive and the operator $L(y)$. (Note: You can use operations in the utility file ODE2 to obtain y_c, y_p, and y_g using a different method. See Example 2.3.)

6. Substitute the initial conditions into y_g and solve for the arbitrary constants using Derive.

7. Plot the solutions to the initial-value problem for $(-4 \leq x \leq 4)$ and the parameter values of $a = -10, 0,$ and 10.

8. Which of the three given parameter values for a produces a maximum at $x = 0$?

9. What is the significance of the parameter in this problem? Relate this model with its parameter to a problem in engineering.

Exercise 3.6

Every body continues in its state of rest, or in uniform motion in a right line, unless it is compelled to change that state by forces impressed upon it.

— Issac Newton

Subject: Vibrations in an Automobile Suspension System

Purpose: Model a physical mechanism with a second-order, nonhomogeneous differential equation, solve the equation, and analyze the behavior.

Given: In the following idealized drawing, an automobile wheel is supported by a spring and shock absorber. On the diagram, $y(t)$ is the vertical displacement from equilibrium ($y = 0$) as a function of time, m is the mass of the car supported by the wheel, and g is the acceleration due to gravity.

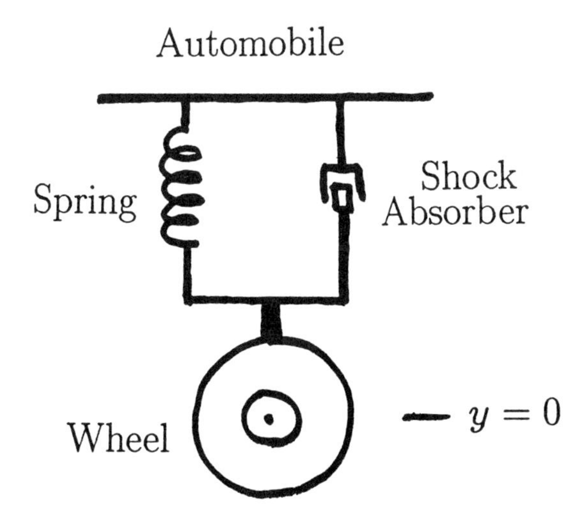

Assume viscous damping in the shock absorber (ie., the frictional force is proportional to the velocity), Hooke's law for the restoring force of the spring, and a forcing function, $f(t)$, which represents the road conditions.

Exercises:

1. Write a second-order, constant coefficient, nonhomogeneous equation for $y(t)$.

2. If a car weighing 3600 lbs. is supported equally by 4 wheels with a spring constant of 200 lbs/ft, a nonnegative resistance coefficient of d, and a forcing function $f(t) = 200 \sin 6t$, write and solve the model for $y(t)$.

3. Assume the further case of no initial vertical displacement or velocity and solve for $y(t)$.

4. Plot $y(t)$ for 3 values of the resistance coefficient of $d = 0$ lb-sec/ft, $d = 70$ lb-sec/ft, and $d = 140$ lb-sec/ft in the interval $0 < t < 4$ seconds.

5. Compare the 3 plots. Which case produces the largest vertical displacement in the interval? Which is the best shock absorber for this road condition?

Exercise 3.7

As is well known, physics became a science only after the invention of differential calculus.

— Riemann [1882]

Subject: Fourth-Order Differential Equation Model for the Deflection of a Beam

Purpose: To solve a higher-order nonhomogeneous differential equation and to analyze the effects on beam deflection of varying parameters in the loading of the beam.

Given: The deflection of a beam carrying a load of $p(x)$ per unit length is modeled with the fourth-order equation

$$EI\frac{d^4w}{dx^4} = p(x),$$

where x is distance along the beam and E and I are positive constants. The physical characteristics of the beam are a length of 30, $E = 3.6$, $I = 2.71$, clamped on the left end ($x = 0$), and free on the right end ($x = 30$). Therefore, the four boundary conditions are $w(0) = 0, w'(0) = 0, w''(30) = 0$, and $w'''(30) = 0$.

Exercises:

1. Find and plot on the same axes $w(x)$ when $p(x) = (2x + a)/100$, for $a = 0, 10$, and 20. What is the total load on the beam for these values of a?

2. Find and plot on the same axes $w(x)$ when

$$p(x) = \begin{cases} 1, & \text{if } \ 0 < x < 10 \\ a, & \text{if } \ 10 < x < 20 \\ 1, & \text{if } \ 20 < x < 30 \end{cases},$$

for $a = 1, 5$, and 10. What is the total load on the beam for the values of a?

3. Find and plot on the same axes $w(x)$ when

$$p(x) = \begin{cases} 1, & \text{if } 0 < x < 20 \\ a, & \text{if } 20 < x < 30 \end{cases},$$

for $a = 1, 5$, and 10. What is the total load on the beam for the values of a?

4. If the beam is supported by an elastic foundation, the model becomes

$$EI\frac{d^4w}{dx^4} + kx = p(x),$$

where k is a positive constant. For a beam with $k = 4$ and the other physical values unchanged from the previous exercises, find and plot on the same axes $w(x)$ when $p(x) = (2x + a)/100$, for $a = 0, 10$, and 20. Compare these deflections with those found in #1. What is the effect of this elastic foundation on the deflection of the beam?

Exercise 3.8

A great discovery solves a great problem but there is a grain of discovery in the solution of any problem.

—George Polya [1946]

Subject: 3 x 3 Homogeneous Linear Differential System with Constant Coefficients

Purpose: Find the solution of a 3 x 3 system of linear, homogeneous equations with an initial value through the determination of eigenvalues and eigenvectors.

Given:

$$x_1' = 3x_1 - 2x_3$$

$$x_2' = -x_1 + x_2 + x_3$$

$$x_3' = 4x_1 - 3x_3$$

with initial conditions

$$\vec{x}(0) = \begin{bmatrix} x_1(0) \\ x_2(0) \\ x_3(0) \end{bmatrix} = \begin{bmatrix} 3 \\ 1 \\ 3 \end{bmatrix}$$

Exercises:

1. Write the system of equations in vector-matrix form.

2. Find the eigenvalues of the characteristic matrix.

3. Find an eigenvector for each of the three eigenvalues.

4. Write the general solution, $\vec{x}_g(t)$ using a, b, and c as the arbitrary constants.

5. Use the initial condition to determine the particular values for a, b, and c.

6. Graph the three components of $\vec{x}(t), x_1, x_2$, and x_3 on one set of axes from $t = 0$ to $t = 10$ (Hint: plot these functions as components of a vector).

Exercise 3.9

I hear and I forget; I see and I remember; I do and I understand.

— Old Chinese Proverb

Subject: Eigenvalues and Eigenvectors

Purpose: Find the eigenvalues and eigenvectors of matrices and use them to determine the general solutions of homogeneous systems of differential equations.

Given: Three 3x3 systems of differential equations of the form $\vec{X}'(t) = \mathbf{A}_j\vec{X}(t)$, with $j = 1, 2, 3$, and

$$\mathbf{A}_1 = \begin{bmatrix} 0 & -1 & -1 \\ 3 & 3/4 & -3/2 \\ -1/2 & 1/8 & 1/4 \end{bmatrix},$$

$$\mathbf{A}_2 = \begin{bmatrix} 2 & 4 & 4 \\ -1 & -2 & 0 \\ -1 & 0 & -2 \end{bmatrix},$$

$$\mathbf{A}_3 = \begin{bmatrix} 2 & 2 & -1 \\ 1 & 0 & 0 \\ 0 & 1 & 0 \end{bmatrix}.$$

Exercises:

1. Find the eigenvalues for the three matrices, $\mathbf{A}_1, \mathbf{A}_2$, and $\mathbf{A}_3$.

2. From analysis of these eigenvalues, which system's solution will decay? Which will grow? Which will show periodic behavior? What is the period of this behavior?

3. Solve for the eigenvectors of matrix $\mathbf{A}_2$ and write the general solution of the differential system in terms of complex numbers and complex variables (i.e. leave the imaginary number i in the eigenvalues and eigenvectors).

4. Simplify the solution in #3 in term of real numbers and real variables.

5. Plot the 3 components of the solution vector for #3 with $\vec{X}(0) = \mathbf{A}_2 = \begin{bmatrix} 1 \\ 1 \\ 1 \end{bmatrix}$ on the same axes in the interval $(-8 < t < 8)$.

Exercise 3.10

If a little knowledge is dangerous, where is the man who has so much as to be out of danger?

— Thomas Huxley [1877]

Subject: Nonhomogeneous System of Differential Equations by Variation of Parameters

Purpose: To solve and analyze a system of three linear differential equations using the method of variation of parameters.

Given:
The following 3x3 differential system with an initial condition:

$$\vec{X}(t)' = \begin{bmatrix} -3 & 2 & 0 \\ 2 & -3 & 0 \\ 0 & 0 & -5 \end{bmatrix} \vec{X}(t) + \begin{bmatrix} 3.25 \\ 6t \\ 2e^t \end{bmatrix},$$

$$\vec{X}(0) = \begin{bmatrix} 0 \\ 1 \\ -1.5 \end{bmatrix},$$

where $\vec{X}(t) = \begin{bmatrix} x(t) \\ y(t) \\ z(t) \end{bmatrix}$.

Exercises:

1. Find the eigenvalues of the matrix for this system.

2. Find an eigenvector for each of the 3 eigenvalues and write the 3 solution vectors for the homogeneous part of the system.

3. Form the fundamental matrix from the set of 3 solution vectors. Use Derive to find a particular solution to the nonhomogeneous system using the formulas in the method of variation of parameters.

4. Find the general solution to the system of equations and evaluate the 3 arbitrary constants from the given initial condition.

5. Plot on the same axes the 3 components of the solution in the region $0 < t < 6$. Which component, x, y, or z, shows the most rapid growth in this region? What is the limiting value for the 3 components as $t \to \infty$?

Exercise 3.11

Education is the instruction of the intellect in the laws of nature.

— Thomas Huxley [1868]

Subject: Linear Models for the Population of Interacting Species

Purpose: Build simple linear system models for the population change of interacting species and determine the solution of the system of differential equations using the help of the mathematical tool Derive.

Given: Populations of indigenous wildlife have developed in an isolated desert park. There are only two known competing species, the jack mole and the prairie rat. Several assumptions have been made in order to build a differential equation model for their populations. $J(t)$ is used for the expression over time of the jack mole population, and $P(t)$ is the expression for the prairie rat. Data collection has indicated both species have identical growth rates which depend only on the size of the population of their own specie and identical competition rates that depend only on the size of the other specie. If $\vec{X}(t) = \begin{bmatrix} J(t) \\ P(t) \end{bmatrix}$, a simple normalized model for this interaction, with the time t measured in years, can be written as

$$\vec{X}'(t) = \begin{bmatrix} 1 & -1 \\ -1 & 1 \end{bmatrix} \vec{X}(t).$$

Exercises:

1. Use Derive to solve for the eigenvalues and eigenvectors of this model. Write the general solution for this system.

2. Without knowing the initial conditions for the populations, determine the possible scenarios for future population changes for these 2 species. Under what condition could both populations survive? What is most likely to happen?

3. If the population sizes are known to be 16,000 jack moles and 15,500 prairie rats, how long will both species coexist?

4. Plot the 2 populations from $t = 0$ until one population no longer exists. Use the movable cross in Derive's plotting window to approximate the time when the smaller population reaches half its original population.

5. If a population of a predator species of coyotes was introduced that preyed on both of the previous species equally at a rate the same as their competition rate and grew at the same rate which depended on the sum of its own population and that of the jack moles, write a new 3x3 matrix model for these 3 populations. Use $C(t)$ to denote the coyote population.

6. Solve this 3x3 model for the general solution, if the coyote population size is 1,000 when the other 2 population sizes are as given in #3.

7. Plot the populations of jack moles and prairie rats for the new model on the same axes as the populations from the previous model. What does the model predict the effect of introducing coyotes into this desert park will be on the other 2 populations?

Exercise 3.12

Modern science ... is an education fitted to promote sound citizenship.

— Karl Pearson [1892]

Subject: Second-order Difference Equation for the National Economy

Purpose: Solve and analyze a difference equation model for the national economy.

Given: The national income during time period n is denoted by $I(n)$ and can be modeled as the sum of incomes of consumer expenditures, private investment, and government expenditures. While these components exist continuously, they are only known and predicted at discrete periods of time. Making assumptions on the behavior of these components over a time period enables a difference equation model to be developed for national income of the form

$$I(n+2) = (1+a)cI(n+1) - acI(n).$$

In this model a is the constant of adjustment, and c is the marginal propensity to consume. Valid values for these two constants are $a > 0$ and $0 < c < 1$.

Exercises:

1. Use commands from the utility file RECUREQN.MTH to solve this model for $a = 1.1$ and $c = 0.3$.

2. If $I(0) = 230$ and $I(1) = 248$, find the solution and plot it for $0 < n < 12$.

3. If $I(0) = 230$ and $I(1) = 270$, find the solution and plot it for $0 < n < 12$ on the same graph as the solution found in #2.

4. What is the value of $i(200)$?

Exercise 3.13

It is better to know some of the questions than all of the answers.

— James Thurber

Subject: Analysis of Inventory and Pricing

Purpose: To solve a nonhomogeneous differential equation that models the pricing policy of a manufacturer.

Given: Based on several assumptions, a manufacturing company has developed simple differential models that relate price change, inventory level, production rate, and sales rate. The pricing policy for an item manufactured by the company is dependent on its inventory level. When the inventory is too high the price decreases and when the inventory is too low the price rises. The simple linear model for this situation is

$$\frac{dp}{dt} = -\mu(L(t) - L_0)$$

where $p(t)$ is the forecast price, $L(t)$ is the inventory level at time t months, L_0 is the desired inventory level as constrained by their warehouse space, and μ is a small (usually ≤ 10), positive constant of proportionality which represents how tightly the inventory level is controlled.

Ultimately, the rate of change of the inventory depends on the production rate $Q(t)$ and the sales rate $S(t)$, which gives the equation

$$\frac{dL}{dt} = Q(t) - S(t).$$

$Q(t)$ and $S(t)$ can be modeled through their dependence on the price and its change by

$$Q(t) = a - bp - c\frac{dp}{dt}$$

and

$$S(t) = \alpha - \beta p - \delta\frac{dp}{dt}$$

where a, b, c, α, β, and δ are constants.

Exercises:

1. Use the given equations to write a second-order differential equation for p in terms of t and the constants.

2. The marketing department has determined the following values for the constants and initial conditions: $a = 0.2, b = 1/4, c = 0.7, \alpha = 14, \beta = 3, \delta = 2, p(0) = 100$, and $p'(0) = 0$. Set up and solve the differential equation in terms of the parameter μ using the appropriate commands from the utility file ODE2.MTH.

3. Plot $p(t)$ for $\mu = 0.1$ and $\mu = 1$ on the same axes showing accurate plots for $0 < t < 3$.

4. Based on the results shown in these plots, determine whether tight inventory control (larger value of μ) or loose control (smaller μ) will result in a lower price for the product at $t = 3$.

5. Use the equation for the sales rate $S(t)$ and the solutions for the two values of μ in #4 to determine which of these two values of μ results in the most total sales over the 3 month period. (Hint: Let Derive do the integration.)

Exercise 3.14

Begin at the beginning ... and go on till you come to the end: then stop.

— Lewis Carroll

Subject: Analysis of a Series Electrical Circuit using Laplace Transforms

Purpose: To solve a second-order, nonhomogeneous model for a simple loop electrical circuit using the techniques of Laplace transforms with Derive performing many of the messy manipulations.

Given: In Section 2.9, a differential model for a simple loop electrical circuit with a time-dependent voltage source ($E(t)$), an inductor (L), a resistor (R), and a capacitor (C) was constructed. Letting $q(t)$ be the time-dependent charge on the capacitor, the model for such a circuit is

$$L\frac{d^2q}{dt^2} + R\frac{dq}{dt} + \frac{1}{C}q = E(t).$$

For this problem the voltage source is defined by $E(t) = \alpha\cos 2t$, for $0 < t < \pi/4$, and 0, thereafter. For a Circuit with $L = 10$ henrys, $R = 50$ ohms, $C = 1/60$ farads, no initial charge on the capacitor ($q(0) = 0$), and no initial current ($q'(0) = 0$), the equation to be solved is

$$10\frac{d^2q}{dt^2} + 50\frac{dq}{dt} + 60q = E(t),$$

with $q(0) = 0$ and $q'(0) = 0$.

Exercises:

1. Write an expression for voltage source $E(t)$ using Derive's unit step function `STEP(t)`.

2. Plot $E(t)$ by accurately showing the interval of $0 < t < 2$ when $\alpha = 110$.

3. Use Derive to find the Laplace transform of $E(t)$. (Note: Either set up the integral definition of a Laplace transform or use the **LAPLACE** command in the utility file INTEGRAL.MTH. Don't forget to establish the proper declaration for the transform variable s.) Set up the algebraic equation resulting from taking the Laplace transform of the differential equation.

4. Solve and expand the algebraic equation and take the inverse Laplace transform using table values to find the solution $q(t)$ to the differential equation.

5. Find the equation for the current $i(t)$ from $q(t)$ using Derive.

6. Plot $i(t)$ for $\alpha = 110$ and $\alpha = -110$. What is the difference between these graphs? Use the movable cross to approximate the maximum absolute value of the current.

Exercise 3.15

Education is what you have left over after you have forgotten everything you have learned.

— Anonymous

Subject: Systems of Equations for the Flow of a Drug through a Biological System

Purpose: To model and analyze drug flow through a biological system using a system of linear differential equations.

Given: A schematic diagram for the fluid flow through a simple biological system made up of Organs A and B is shown.

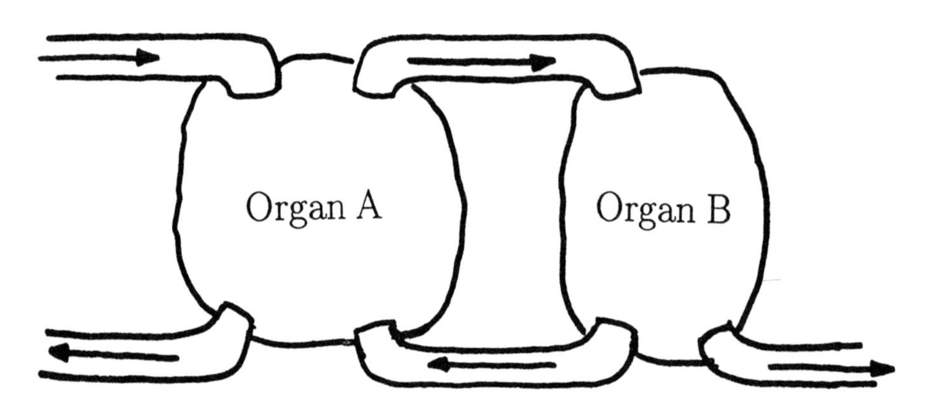

Organ A stays filled with 3 gallons of fluid and Organ B maintains 2 gallons. The following fluid flows are established: external input to A is 0.05 gal/hour, external output from A is 0.03 gal/hour, flow from A to B is 0.05 gal/hour, flow from B to A is 0.03 gal/hour, and the external output of B is 0.02 gal/hour.

Exercises:

1. If the input into A is pure fluid (no drugs), write the differential equation model for the amount of the drug in Organ A (x_1) and Organ B (x_2).

2. If 3 oz of the drug are injected directly into Organ A and Organ B is free of the drug, solve the model for x_1 and x_2. Plot the equations for $0 < t < 80$.

3. When will the drug concentration in organ B exceed 0.1 oz/gal and when will it fall below 0.1 oz/gal?

4. Model and solve for x_1 and x_2 if the 3 oz of the drug are injected directly into Organ B instead of Organ A. Plot the components of the solution. When will the drug concentration in organ B fall below 0.1 oz/gal for these initial conditions?

5. Instead of injecting the drug all at once, it is introduced into the system through the input flow into A at a concentration of 6 oz/gal. Write the new model for this flow. How much of the drug is introduced into the system after 10 hours?

6. Solve the model in #5. Plot the solution. When will the drug concentration in Organ B exceed 0.1 oz/gal and fall below 0.1 oz/gal for this model?

Exercise 3.16

...it behooves us to place the foundations of knowledge in mathematics.

— Roger Bacon [13th century]

Subject: Analysis of a Series Electrical Circuit using Systems of Differential Equations

Purpose: To solve a nonhomogeneous system of two differential equations that model a two loop electrical circuit.

Given: In Section 2.9, a differential model for a simple loop electrical circuit with a voltage source ($E(t)$), an inductor (L), a resistor (R), and a capacitor (C) was constructed. Now a circuit with two loops in series needs to be modeled and analyzed. The schematic diagram for the Circuit is given below.

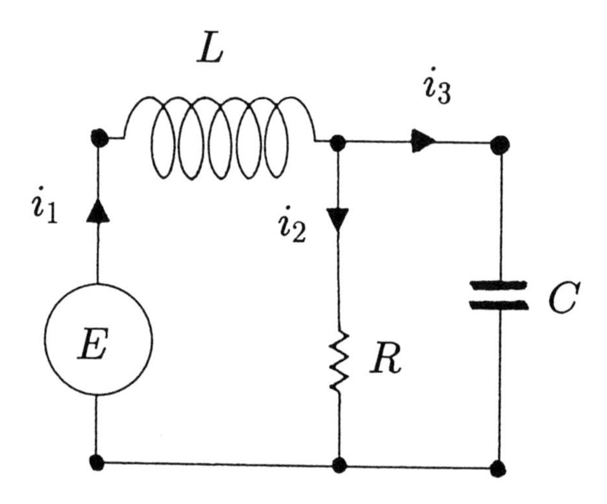

Letting $i_k(t), k = 1, 2$, and 3 be the time-dependent current for the wires in the loops as shown, the model for this circuit is

$$L\frac{di_1}{dt} + Ri_2 = E(t),$$

$$RC\frac{di_2}{dt} + i_2 - i_1 = 0.$$

For this problem the voltage source is defined as a constant $E(t) = \alpha$ volts, $L = 0.5$ henrys, $R = 50$ ohms, $C = 0.0001$ farads, $i_1(0) = 0$ amps, and $i_2(0) = 0$ amps.

Exercises:

1. Write the model as a matrix-vector system. Determine the eigenvalues, eigenvectors, and complementary solution for the homogeneous system.

2. Solve the nonhomogeneous system using variation of parameters (Section 2.8) subject to the given boundary conditions.

3. If i_1 cannot exceed 2 amps, what is the maximum value for α?

4. Plot $i_1(t)$ and $i_2(t)$ on the same axes. What is the long-term (steady state) behavior of these two currents?

Exercise 3.17

There is in mathematics hardly a single infinite series of which the sum is determined in a rigorous way.

—Niels Henrik Abel [1826]

Subject: Approximating Functions with Fourier Series

Purpose: Use Derive to find terms of the Fourier series for several functions using the function command in the Utility file INTEGRAL.MTH (See Example 2.11).

Given: The following functions, $f_i(t)$:

$$f_1(x) = \begin{cases} 0, & \text{if } -\pi \le x < 0 \\ 1 & \text{if } 0 \le x \le \pi \end{cases}$$

$$f_2(x) = \begin{cases} -x & \text{if } -2 \le x < 0 \\ x & \text{if } 0 \le x \le 2 \end{cases}$$

$$f_3(x) = \begin{cases} 0 & \text{if } -3 \le x < -1 \\ 1 & \text{if } -1 \le x < 1 \\ 0 & \text{if } 1 \le x \le 3 \end{cases}$$

$$f_4(x) = \begin{cases} 0 & \text{if } -\pi \le x < 0 \\ x & \text{if } 0 \le x \le \pi \end{cases}$$

$$f_5(x) = -x \quad \text{if} \quad -\pi < x \le 0$$

$$f_6(x) = e^x \quad \text{if} \quad -\pi < x \le \pi$$

$$f_7(x) = \cos(x^2)e^x \quad \text{if} \quad -\pi < x \le \pi$$

Exercises:

1. Load the Utility file INTEGRAL.MTH into the work space. Use the function **FOURIER** to attempt to find the first three terms of the Fourier series for these 7 functions. Which functions can't it evaluate? If the evaluation cannot be performed in the **Exact** mode, change Derive to the **Approximate** mode. Does any calculation give a warning of 'Dubious Accuracy'?

2. Plot each of the 7 functions and its 3-term approximation in the interval $-8 < x < 8$. What are the periods of these functions? Which functions are even or odd?

3. Find the first 6 terms for each of these 7 functions. Plot the 3-term and the 6-term approximation for each of the functions on the interval $-8 < x < 8$. Does the approximation improve with more terms?

Exercise 3.18

The purpose of computing is insight, not numbers.

—R. W. Hamming

Subject: Evaluating Laplace Transforms

Purpose: Use Derive to find the Laplace transform of several types of functions using the integral definition in the Utility file INTEGRAL.MTH (See Example 2.18).

Given: The following functions, $f_i(t)$:

$$f_1(t) = \sin 3t \cos 3t$$

$$f_2(t) = \begin{cases} -1, & \text{if } 0 \le t < 2 \\ 1, & \text{if } t \ge 2 \end{cases}$$

$$f_3(t) = \sin^2 t$$

$$f_4(t) = e^{-t} \sin^2 t$$

$$f_5(t) = \begin{cases} \cos t, & \text{if } 0 \le t < \pi \\ t, & \text{if } t \ge \pi \end{cases}$$

$$f_6(t) = e^{-t} t \cos 2t$$

$$f_7(t) = \frac{\sin 3t}{t}$$

Exercises:

1. Plot each of the 7 functions in the interval $(0 < t < 10)$. Think of them as forcing functions of a mechanical system. Which are periodic, discontinuous, smooth?

2. Load the Utility file INTEGRAL.MTH into the work space. Use the function **LAPLACE**(u, t, s) to attempt to evaluate the integral definition of the Laplace transform of these 7 functions. Which functions can't it evaluate? Can you do the transform quicker by hand using a table?

3. From this test case of 7 functions, is it worth your time to use Derive to do Laplace transforms?

Exercise 3.19

A traveler who refuses to pass over a bridge until he has personally tested the soundness of every part of it is not likely to go far; something must be risked, even in mathematics.

— Horace Lamb

Subject: Separation of Variables for the Heat Equation

Purpose: To solve the heat equation with nonhomogeneous boundary conditions and to investigate the effects of varying parameters in the initial and boundary conditions.

Given: The temperature $u(x,t)$ of a rod of length 15 is modeled by the following heat equation:

$$ku_{xx} = u_t, \quad 0 < x < 15, \quad t > 0,$$

$$u(0,t) = a, \quad u(15,t) = 0, \quad u(x,0) = b,$$

where k is a constant thermal diffusivity and a and b are parameter values for temperature.

Exercises:

1. Use the change of dependent variable ($u(x,t) = v(x,t) + w(x)$) to transform the given equation into a heat equation for v with homogeneous boundary conditions and into a second-order ordinary differential equation for w with nonhomogeneous boundary conditions. The two parts of the solution v and w are called the transient and steady-state solutions, respectively.

2. Solve this heat equation for v and the differential equation for w. See Example 2.13. Does the steady-state solution depend on k, a, or b? Explain why.

3. Find the expression for the temperature distribution $u(x,t)$.

4. If $k = 0.853$ (assume this value for k for the rest of the problem), use the first term in the series for the transient solution and the steady-state solution to approximate the temperature at $x = 6$ when $t = 1, 10, 20$, and 40.

5. What are the values computed in #4 when $a = 100$ and $b = 100$? Use 2 terms of the series to approximate the same values.

6. If $a = 100$ and $b = 100$, plot on the same axes the 3-term approximations of the temperature distributions for $t = 0, 10$, and 20 and the steady-state solution. How close is the temperature distribution to steady state by $t = 20$?

7. If $a = 100$, plot on the same axes the 3-term approximations of the temperature distribution for $t = 5$ when $b = 0, 10, 20$, and 40. Does the initial temperature have a lasting effect on the temperature distribution?

8. If $b = 100$, plot on the same axes the 3-term approximation of the temperature distribution for $t = 5$ when $a = 0, 10, 20$, and 40. Does the value of the boundary condition parameter a have a lasting effect on the temperature distribution?

Exercise 3.20

...the different branches of Arithmetic—Ambition, Distraction, Uglification, and Derision—

— Lewis Carroll in *Alice in Wonderland* [1865]

Subject: Solving Linear and Bernoulli Equations

Purpose: To solve and analyze linear and Bernoulli equations.

Given: The commands `LINEAR1` and `BERNOULLI` are provided in the utility file ODE1.MTH to solve initial-value problems of special forms. There are brief descriptions of these commands in Section 1.12 and in the file ODE1.MTH.

Exercises:

1. Solve and plot the solution of

$$xy' + (1+x)y = e^{-x}\sin(2x), \quad y(1) = \sqrt{2}.$$

What is the limiting value of the solution as $x \to \infty$?

2. Solve and plot the solution of

$$y' + (\tan x)y = \cos^2 x, \quad y(\pi) = 2.$$

3. Solve and plot on the same axes the solutions of

$$y' + y = xy^n, y(0) = 1,$$

for $n = 2$ and 3.

4. Solve and plot on the same axes the solutions of

$$xy' + y = 2x, y(a) = b,$$

for the following four pairs of values for a and b: $(a, b) = (-2, 0)$, $(-1, -2)$, $(1, 2), (2, 0)$.

Index